Industrial Robots:
Application Experience

Joint Editors: Prof. Dr.-Ing. H. J. Warnecke
Dr. R. D. Schraft

I.F.S. Publications Ltd.
35-39 High Street, Kempston, Bedford MK42 7BT, England

DISTRIBUTED BY
SCHOLIUM
INTERNATIONAL, INC.
265 GREAT NECK ROAD
GREAT NECK, NEW YORK 11021

British Library Cataloguing in Publication Data

Warnecke, H. J.
  Industrial robots: application experience.
  1. Robots, Industrial
  I. Title II. Schraft, R. D.
  III. Industrieroboter. *English*
  629.8′92   TS191

  ISBN 0-903608-21-9

Typesetting by Fotographics (Bedford) Limited; printed by Cotswold Press, Oxford, U.K.

Contents

v

1. Introduction

Six years have elapsed since the publication of the first edition of the book "Industrieroboter" (Industrial Robots) (1.1). During this period, considerable efforts have been made on the one hand to develop the units further, and on the other to solve the problems relating to their use. At that time the industrial robot was regarded primarily as a means of automation which would solve technical and economic problems in the handling of rapidly changing tasks. In the meantime the factors leading to the use of industrial robots have increased due to the efforts being made to humanise the working environment. These machines are thus no longer solely means for automating work processes, but they are also used in work situations where workers are subjected to severe physical and/or psychological stress. The absence of flexible, peripheral equipment or of suitable sensor systems, which at that time prevented the use of these machines in many cases has led in the intervening period to a great variety of research and development activities with the result that today a considerably greater range of suitable peripheral equipment (magazines, feed devices, etc.) is available. However, these problems have by no means been solved entirely satisfactorily as can be seen by the worldwide activities in this field. The large number of suppliers who at that time frequently had only prototypes to offer has since contracted. Firms which were not able to offer complete solutions to the problems have in most cases disappeared from the market, whilst new firms have been established which are able to offer the user a variety of solutions to his problems, including industrial robots.

1.1   Explanation of the term "robot"

The word "robot" is derived from the Slav word "robota", meaning "heavy work". In general usage, the term "robot" is understood to mean a humanoid machine which, under certain circumstances, resembles the appearance of the human being or which is able to perform human functions, at least partially (1.1). Asimov (quoted in (1.2)) has even formulated laws of "robotics" which govern the behaviour of "robots" towards human beings. In the technical literature, "robots" are understood to be machines which have the capability of performing independently tasks of a mental and physical nature (1.3, 1.4, 1.5, 1.6, 1.7, 1.8, 1.9). To overcome the problems encountered in this field, they must interact independently and "to some extent intelligently" with their environment, and must be capable of storing experience once it has been gained.

In order to build "robots" on the principle just outlined it must first be clearly established which characteristics and capabilities the human being possesses in

1

order to reproduce them technically. According to Thring (quoted in (1.2)), a "robot" must possess the following elements and capabilities:

1. An arm and a hand
2. A facility for travelling and steering
3. A drive and control system for 1. and 2.
4. A minicomputer for sorting commands and for decision making.
5. Sensors for contact, roughness, hardness, position, weight, thermal conductivity, temperature, proximity, shape, size, appearance, colour, distance, odour, position of the arm and hand, hearing.

This list of requirements for "robots" is based largely on the characteristics and capabilities of man. Steinbruch (1.10) has investigated the cybernetic characteristics and capabilities of man, and the result is summarised in Figure 1.1.

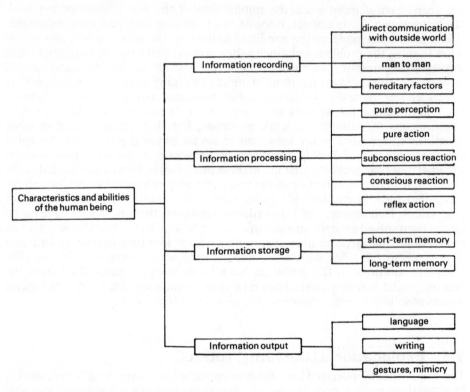

Figure 1.1   Characteristics and capabilities of man, according to Steinbruch (1.10).

1.2   Technical feasibility

If we analyse the list of requirements, according to Thring, for "robots", in section 1.1, it would appear that the requirements for mainly physical characteristics 1. to 3. can be provided by the technical means now available. Requirements 4. and 5. are related to information processing and are largely covered by the characteristics and capabilities of man shown in Figure 1.1.

In the following we shall examine how far these characteristics and capabilities can now be simulated technically.

1.2.1 Information recording
Technically, information recording takes place by means of sensors. Sensors may be divided into non-contact and contact (tactile) sensors (see also Chapter 6).

1.2.2 Information processing
The sensors for information recording provide a "picture of the environment", which must be evaluated and processed. This is done by a comparison with information already stored by means of a classification process. In the literature, this type of information processing is termed "pattern recognition", which may, for example, be effected by methods of scene analysis (1.11).

The process of pattern recognition consists of three individual operations:
(a) separation of objects from the background.
(b) establishment of characteristic properties of the object, such as shape, size colour, etc.
(c) comparison of the properties determined with properties of known objects.
The comparison mentioned under c. may be carried out in two different ways:
1. By a congruence comparison (correlation).
    These methods are considerably limited in their information content. The recognition of an object is in this case optical/geometrical (e.g. by using masks) and is therefore limited to the shape of the object. Moreover, the patterns to be recognised must be standardised, i.e. they must coincide with one another and with the mask, e.g. when selecting required workpieces by means of a correlator (1.12).
2. By hierarchial comparison of the established properties with the stored properties of known objects.
These methods provide much more information on the objects, as they are able to determine not only shape, but also colour, location and size. They examine objects progressively and hierarchically for their characteristics. The concentration of information increases with every stage until a clear recognition of the object is guaranteed.

1.2.3 Information storage
The task of information storage consists in holding information that has already been collected and making it available when required. Nothing definite is known about the biochemical processes which occur in man during information storage in the brain (1.13). Technical information stores and memories abound. Examples of well-known storage systems for digital information processing are code memories, magnetic discs, punched tapes, etc. Chapter 2.3).

1.2.4 Information output
The necessity for outputting information arises from the need to communicate with the surrounding world. Technical systems are also able to communicate with their surroundings. For example, technical possibilities for the exchange and transmission of information include acoustic signals, optical signals, recorders, printers, screens, etc.

1.3 State of the art

At a number of research institutes and industrial firms development work is being done on sensors already descibed. Manipulators are also being developed which possess a modest amount of artificial intelligence.

The following examples illustrate this:

The firm Hitachi Ltd., Japan, has completed a development in which a manipulator automatically performs simple assembly operations by means of tactile sensors in the gripper and a computer. In this case, the device searches for workpieces in a working area known to it, recognises or senses them by means of its sensors, and arranges them in a prepared box (1.14). At the Stanford Research Institute, USA, a unit has been developed which is able to move automatically in an environment known to it, and which is capable of bypassing unforeseen obstructions or of moving them out of the way. The device is equipped with a number of sensors with which to recognise the environment or obstructions and to localise them. Its travel movements are stored on the basis of speed and direction, so that the device always knows its present position. These measurements are susceptible to errors, as with all technical installations, and the calculated positions are therefore compared at regular intervals with a television picture, which is recorded, for example, from one corner of the known area. Both these sets of information are evaluated in order to re-determine the position of the device, if necessary.

These examples show that attempts are being made to find technical solutions for the requirements 4. and 5. by Thring. However, these developments are still so expensive that they cannot be considered for industrial application for the present (for further examples see also Chapter 6).

1.4 Equipment for industrial application

The majority of devices which are now being constructed and used industrially may be divided into three groups, viz:
- pick-and-place devices
- industrial robots
- tele-operators

Figure 1.2
Pick-and-place unit.
(Photograph: Felss).

All these units require a predetermined environment, since they have no, or only very simple, sensors.

Pick-and-place units have been very well known for some time and are used in

4

different branches of industry. In most cases they are mechnical handling devices equipped with grippers, which perform predetermined movements according to a fixed programme (1.1). They operate on machine tools, assembly lines, in the packing industry, etc., in fact wherever the same handling operation (1.15, 1.16) has to be carried out repeatedly over a long period. A typical device is shown in Figure 1.2.

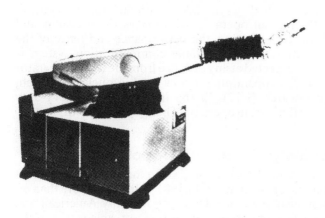

Figure 1.3   Industrial robot. (Photograph Unimation Inc.).

Figure 1.4:   Tele-operator "Man-Mate" (Photograph: General Electric).

Industrial robots are different from these; in most cases they are described in the relevant literature (1.3, 1.4, 1.5, 1.6, 1.7, 1.8, 1.9) as programmable handling devices. They are automatic handling devices freely programmable in several axes of movement, equipped with grippers or tools, and are designed for industrial use (1.1). The difference between these devices and pick-and-place units lies in their programmability and their, in most cases, more complicated kinematics. Figure 1.3 shows an example of an industrial robot.

Tele-operators are remote-controlled manipulators without program control. This task is performed by the human operator, who makes the decisions and initiates the movements. With tele-operators, the performance and range of the human operator can be far surpassed. If a suitable communication system is available, the tele-operator can be erected at any distance from the human operator and can operate at that point. Tele-operators are now used in nuclear research, marine research and space research (1.17). In the industrial field, hevy-duty manipulators are used to relieve the human operator of heavy physical work.

1.5 Present development of industrial robots

The development of industrial robots began in the USA in the mid 1960s. In 1968 these devices appeared on the European market. In 1981 there were more than 50 suppliers in Europe selling industrial robots. There are not only American and Japanese, but also European developments, and what is remarkable is that in the last few years a whole series of new developments have taken place in Europe. It should also be noted that compared with 1973, only very few Japanese devices, which were being offered at that time, are now available on the European market.

2. Features of industrial robots

2.1 Sub-systems and functional structure of robots

An industrial robot is a complex, technical system, consisting of several sub-systems. Each of these sub-systems performs defined subsidiary functions, and therefore contributes towards fulfilling the overall function of the industrial robot. The breakdown into subsidiaries and sub-systems gives an insight into the internal structure of the system. Figure 2.1 illustrates the systems which are found, either singly or in multiples, in an industrial robot, together with the associated functions.

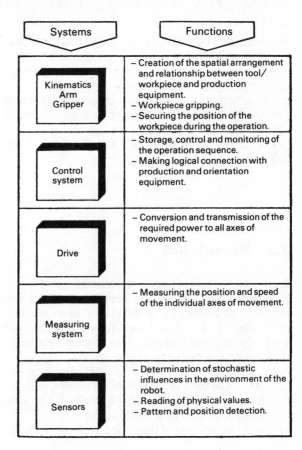

Systems	Functions
Kinematics Arm Gripper	– Creation of the spatial arrangement and relationship between tool/workpiece and production equipment. – Workpiece gripping. – Securing the position of the workpiece during the operation.
Control system	– Storage, control and monitoring of the operation sequence. – Making logical connection with production and orientation equipment.
Drive	– Conversion and transmission of the required power to all axes of movement.
Measuring system	– Measuring the position and speed of the individual axes of movement.
Sensors	– Determination of stochastic influences in the environment of the robot. – Reading of physical values. – Pattern and position detection.

Figure 2.1
Systems and functions
of an industrial robot.

7

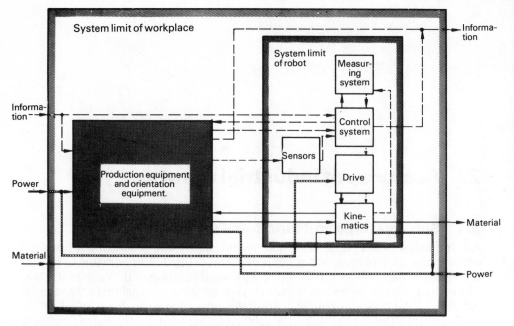

Figure 2.2 Functional structure of a workplace with industrial robot.

The functional structure of an industrial robot is represented in Figure 2.2, which shows the interlinking of the sub-systems by the material, information and energy flows and their connection with the working environment and the production and orientation systems.

The kinematics is described in detail in Section 2.2, and the control system is described in Section 2.3. Chapter 6 describes not only the development trends but also the state of the art in sensor engineering.

The sub-systems with their functions are described in detail in the following sections.

2.2 Kinematics

2.2.1 Definition of axes of movement

Kinematics are understood here to mean the spatial arrangement, according to sequence and structure, of the axes of movement (degrees of freedom) in relation to each other. The task of kinematics is to enable arbitrary spatial points in a work area to be approached and there to create the desired spatial relationship between the gripper or tool and this point.

No definite agreement has yet been reached in VDI committees on the designation of the axes of industrial robots, and until that is done the notation used up till now will be continued (Figure 2.3). The kinematic description of completed machines is divided into three blocks:

1st block: movement of the entire machine (movement of travel).

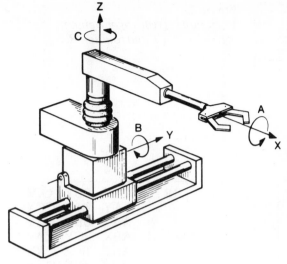

Figure 2.3 Modular description of the movement of an industrial robot. (Photograph: Telles).

2nd block: movement of the arm.
3rd block: movement of the gripper.
    An example of a block description is shown in Figure 2.3.

2.2.2 Variations and combinations of axes of movement

On examining all the possibilities of combining three rotary and/or translatory axes of movement, i.e. the forming of kinematic chains, $2^3 = 8$ possible variations are obtained (Figure 2.4).

    If the axes X, Y, Z, A, B, C are combined for such a variation, $3^3 = 27$ combinations are obtained.

    The total number of all possible combinations of movements to form kinematic chains is obtained by multiplying the variants and combinations:
8 variants multiplied by 27 combinations = 26 kinematic chains.

    But some of these 216 kinematic chains (Figure 2.5), formed from theoretical considerations, do not produce a swept volume. It is assumed here that all axes of rotation possess arms. The analysis results in the following relations:
(a) Operation of the X, Y, C axes leads to movements in plane X, Y
(b) Operation of the Z, Y, A axes leads to movements in plane Z, Y
(c) Operation of the Z, X, B axes leads to movements in plane Z, X
This means that no volume is swept when:
 (i) axes combine in one plane
(ii) two identical axes of translation combine
    Now if we consider all 216 kinematic chains in relation to these criteria, 87 kinematic chains can be found which only move in a single plane or on the surface of a solid. Thus in the following only 129 kinematic chains (Figure 2.6) will be considered.

    Since some kinematic chains are only distinguished by their notation and others can be converted into chains with a different notation by rotation, it is necessary to formulate transformation rules (Figure 2.7) to discover these.

	R	T
RR	RRR	RRT
RT	RTR	RTT
TR	TRR	TRT
TT	TTR	TTT

Figure 2.4
Variations of types of movement of translation T and rotation R

RRR	AAA BBB CCC
	AAB BBC CCA
	AAC BBA CCB
	ABA BCB CAC
	ABB BCC CAA
	ABC BCA CAB
	ACA BAB CBC
	ACB BAC CBA
	ACC BAA CBB
RTR	AXA BYB CZC
	AXB BYC CZA
	AXC BYA CZB
	AYA BZB CXC
	AYB BZC CXA
	AYC BZA CXB
	AZA BXB CYC
	AZB BXC CYA
	AZC BXA CYB
TRR	XAA YBB ZCC
	XAB YBC ZCA
	XAC YBA ZCB
	XBA YCB ZAC
	XBB YCC ZAA
	XBC YCA ZAB
	XCA YAB ZBC
	XCB YAC ZBA
	XCC YAA ZBB
TTR	XXA YYB ZZC
	XXB YYC ZZA
	XXC YYA ZZB
	XYA YZB ZXC
	XYB YZC ZXA
	XYC YZA ZXB
	XZA YXB ZYC
	XZB YXC ZYA
	XZC YXA ZYB

RRT	AAX BBY CCZ
	AAY BBZ CCX
	AAZ BBX CCY
	ABX BCY CAZ
	ABY BCZ CAX
	ABZ BCX CAY
	ACX BAY CBZ
	ACY BAZ CBX
	ACZ BAX CBY
RTT	AXX BYY CZZ
	AXY BYZ CZX
	AXZ BYX CZY
	AYX BZY CXZ
	AYY BZZ CXX
	AYZ BZX CXY
	AZX BXY CYZ
	AZY BXZ CYX
	AZZ BXX CYY
TRT	XAX YBY ZCZ
	XAY YBZ ZCX
	XAZ YBX ZCY
	XBX YCY ZAZ
	XBY YCZ ZAX
	XBZ YCX ZAY
	XCX YAY ZBZ
	XCY YAZ ZBX
	XCZ YAX ZBY
TTT	XXX YYY ZZZ
	XXY YYZ ZZX
	XXZ YYX ZZY
	XYX YZY ZXZ
	XYY YZZ ZXX
	XYZ YZX ZXY
	XZX YXY ZYZ
	XZY YXZ ZYX
	XZZ YXX ZYY

Figure 2.5   Mathematically possible kinematic chains from the movement axes A, B, C, D, X, Y, Z.

10

RRR	AAB BBC CCA	TRR	XAA YBB ZCC
	AAC BBA CCB		XAB YBC ZCA
	ABA BCB CAC		XAC YBA ZCB
	ABB BCC CAA		XBA YCB ZAC
	ABC BCA CAB		XBC YCA ZAB
	ACA BAB CBC		XCA YAB ZBC
	ACB BAC CBA		XCB YAC ZBA
	ACC BAA CBB		
RRT	AAX BBY CCZ	TRT	XAY YBZ ZCX
	ABX BCY CAZ		XAZ YBX ZCY
	ABY BCZ CAX		XBY YCZ ZAX
	ABZ BCX CAY		XCZ YAX ZBY
	ACX BAY CBZ		
	ACY BAZ CBX		
	ACZ BAX CBY		
RTR	AXA BYB CZC	TTR	XYA YZB ZXC
	AXB BYC CZA		XYB YZC ZXA
	AXC BYA CZB		XZA YXB ZYC
	AYB BZC CXA		XZC YXA ZYB
	AYC BZA CXB		
	AZB BXC CYA		
	AZC BXA CYB		
RRT	AXY BYZ CZX	TTT	XYZ YZX ZXY
	AXZ BYX CZY		XZY YXZ ZYX
	AYX BZY CXZ		
	AZX BXY CYZ		

Figure 2.6   Kinematic chains sweeping a volume.

End movement after rotation about \ Initial movement	A	B	C	X	Y	Z
u-axis	A	C	B	X	Z	Y
v-axis	C	B	A	Z	Y	X
w-axis	B	A	C	Y	X	Z

*Figure 2.7   Transformation rules for rotation about the spatially fixed u-, v- and
w-axes.*

11

Initial kinematics		Rotation about the axes					
		X	Y	Z	X	Y	Z
RRR	AAB	AAB	CCB	CCB	BBC	BBC	AAB
	AAC	AAC	CCA	CCA	BBA	BBA	AAC
	ABA	ABA	CBA	CBA	BCA	BCA	ABA
	ABB	ABB	CBB	CBB	BCC	BCC	ABB
	ABC	ABC	CBC	CBC	BCB	BCB	ABC
	ACA	ACA	CAC	CAC	BAC	BAC	ACA
	ACB	ACB	CAB	CAB	BAB	BAB	ACB
	ACC	ACC	CAA	CAA	BAA	BAA	ACC
RRT	AAX	AAX	CCZ	CCZ	BBY	BBY	AAX
	ABX	ABX	CBX	CBX	BCX	BCX	ABX
	ABY	ABY	CBY	CBY	BCZ	BCZ	ABY
	ABZ	ABZ	CBZ	CBZ	BCY	BCY	ABZ
	ACX	ACX	CAZ	CAZ	BAZ	BAZ	ACX
	ACY	ACY	CAY	CAY	BAY	BAY	ACY
	ACZ	ACZ	CAX	CAX	BAX	BAX	ACZ
RTR	AXA	AXA	CZC	CZC	BYB	BYB	AXA
	AXB	AXB	CZB	CZB	BYC	BYC	AXB
	AXC	AXC	CZA	CZA	BYA	BYA	AXC
	AYB	AYB	CYB	CYB	BZC	BZC	AYB
	AYC	AYC	CYA	CYA	BZA	BZA	AYC
	AZB	AZB	CXB	CXB	BXC	BXC	AZB
	AZC	AZC	CXA	CXA	BXA	BXA	AZC
RTT	AXY	AXY	CZY	CZY	BYZ	BYZ	AXY
	AXZ	AXZ	CZX	CZX	BYX	BYX	AXZ
	AYX	AYX	CYZ	CYZ	BZY	BZY	AYX
	AZX	AZX	CXZ	CXZ	BXY	BXY	AZX
TRR	XAA	XAA	ZCC	ZCC	YBB	YBB	XAA
	XAB	XAB	ZCB	ZCB	YBC	YBC	XAB
	XAC	XAC	ZCA	ZCA	YBA	YBA	XAC
	XBA	XCA	ZAC	ZBA	YCA	YAC	XBA
	XBC	XCB	ZAB	ZBC	YCB	YAB	XBC
TRT	XAY	XAY	ZCY	ZCY	YBZ	YBZ	XAY
	XAZ	XAZ	ZCX	ZCX	YBX	YBX	XAZ
	XBY	XCZ	ZAX	ZBY	YCZ	YAX	XBY
TTR	XYA	XZA	ZXC	ZYC	YZB	YXB	XYA
	XYB	XZC	ZXA	ZYB	YZC	YXA	XYB
TTT	XYZ	XZY	ZXY	ZYX	YZX	YXZ	XYZ

Figure 2.8   *Changes in the designation of the kinematic chains by rotation (Σ37 kinematic chains).*

Figure 2.9
Transformation
rules
for kinematics.

Axis of Rotation	Initial Movement	Final Movement	Succeeding axis of rotation
u-axis	A	A	u-axis
	B	C	-
	C	B	-
	X	X	-
	Y	Z	-
	Z	Y	-
v-axis	A	C	-
	B	B	-
	C	A	-
	X	Z	-
	Y	Y	-
	Z	X	-
w-axis	A	B	v-axis
	B	A	u-axis
	C	C	w-axis
	X	Y	-
	Y	X	-
	Z	Z	-

Movement A is defined as rotation about the u-axis: consequently movement A, with rotation about the u-axis, still satisfies the definition conditions of a movement A, whilst, after the same rotation a movement B revolves about the fixed w-axis, and is therefore defined as movement C. In addition to these considerations, further limitations must be made with regard to the structural design of industrial robots.

Axis A consists of a bearing, which permits rotation about the horizontal u-axis, and of a lever arm of defined length perpendicular to this axis. The lever arm is aligned in direction w. Similarly, a movement C and B is associated with a lever arm of defined length aligned in direction u.

When establishing the rules of transformation, the lever arms of the axes of rotation must be considered as having a defined alignment. An A movement, when rotating about the w-axis, produces movement B, for example, and the lever arm must be aligned in direction u, which is equivalent to a rotation of all succeeding axes about the v-axis. Thus when these definitions are included, transformation rules according to Figure 2.9 are obtained.

On examining the 129 kinematic chains from figure 2.6, with the transformation rules from Figure 2.7, only 37 different kinematic chains are then obtained (Figure 2.8). The structure of these kinematic chains is shown in Figure 2.10.

Though these 37 possible kinematic systems, which produce a working space, are possible in principle, in practice 4 basic designs have prevailed. These are shown in Figures 2.11 to 2.14.

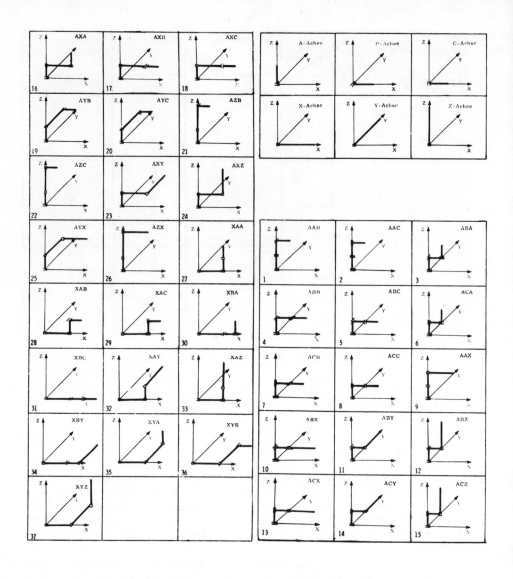

*Figure 2.10   Definition of the kinematic axes and representation of the different
kinematic chains.*

14

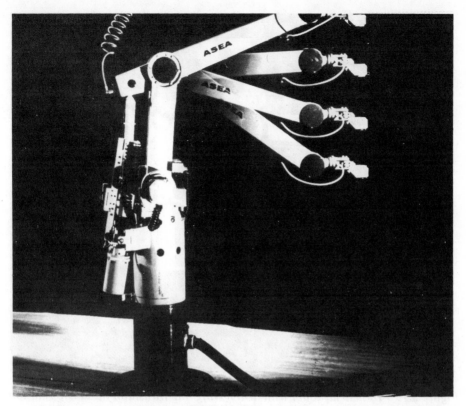

Figure 2.11   Industrial robot with arm having three rotational movements.
(Photograph: ASEA).

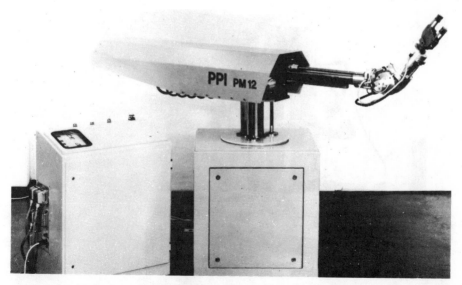

Figure 2.12   Industrial robot with arm having two translatory and one rotational
movement. (Photograph: PPI GmbH).

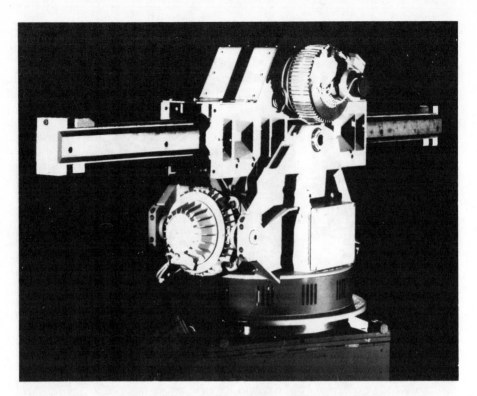

Figure 2.13   Industrial robot with arm having one translatory and two rotational movements. (Photograph: VW AG).

Figure 2.14   Industrial robot with arm having three translatory movements (Photograph: Kuka).

2.3 Control systems

The flexibility and efficiency of an industrial robot are determined by its control system, within the limits set beforehand by the design of the mechanical structure.

The task of the control system is primarily to provide a logical sequence for the operating program. It provides the theoretical position values required for each program step, continuously measures the actual position during the movement, and processes the theoretical/actual difference, together with other measured values (travelling speeds) and stored data (e.g. theoretical speeds, dwell times) into actuating variables for the drives.

The program is generally also produced with the help of the control system, but programming methods are also conceivable where the programs are executed independently of the control system at a particular programming point.

Electrical control systems are generally used, and pneumatic control systems are only used for simple, pneumatically driven industrial robots and pick-and-place units. For research purposes more complex pneumatic control systems have been developed, the use of which might be feasible under special environmental conditions (e.g. explosion risk), though practical applications are not known (2.1, 2.2, 2.3).

By contrast to numerical control engineering, it is not common practice to use industrial robots and control systems from different manufacturers and in this way to adapt properties and "intelligence" separately to a planned application.

Due to the lack of standardised interfaces and the large number of possible kinematic designs, drives and movement measuring systems, this is not possible in many cases, or is only possible as a result of extensive adaptations.

Since the introduction of mini- or micro-computers as central processing units for the control systems, their flexibility has also increased considerably, so that the cost of adapting a different control system is in most cases disproportionate to the gain in efficiency.

Generally speaking intervention by the user in the control system, or even its replacement, is only necessary in those cases where several industrial robots are operated by a supervisory control system (2.4).

The standardisation of interfaces in such cases is now being discussed by the competent bodies.

2.3.1 Types of control systems

A basic distinction is made between two types of control systems for industrial robots: the point-to-point control system, and the continuous path control system.

2.3.1.1 Point-to-point control systems

Most of the units now on the market are equipped with a point-to-point control system. Such a control system can be used in all cases where only individual points within the working area have to be approached, for handling operations, for example, such as the operation of injection moulding or die casting machines, machine tools, presses, the loading and unloading of magazines or pallets, or the handling of spot welding electrode holders.

During the movement, each axis approaches the programmed theoretical co-ordinates and there is no functional relationship between the individual axes.

The path of movement of the gripper is not defined and is difficult to predict, particularly in the case of machines with a system of non-Cartesian co-ordinates, as far as the programmer is concerned.

It depends not only on the initial and end points, but also on the programmed speed and the load. If certain obstacles have to be bypassed on the way from the starting to the end points, intermediate points must be programmed. In many control systems such points may be characterised in the program separately as auxiliary points. The movement is not then interrupted at these points. Many manufacturers use the term multi-point control if an exceptionally large number of points can be stored. If these points lie close together and if the movement is not slowed down at each point but instead becomes a continuous path by passing by each of the points, the behaviour of a continuous path control system can be approximated by using such a system. However, programming is troublesome because it must be carried out manually, point by point.

2.3.1.2 Continuous path control systems

Continuous path control systems are used wherever a certain path must be reproduced independently, as far as possible, of the speed of movement and load of the industrial robot. Common applications are painting and coating, parting-off by abrasive disc and seam welding. Inside the control system, the path to be travelled is represented by a large number of points in close proximity. The co-ordinates of these points are either measured and stored during actual programming, or they are calculated during the movement by interpolating between points further apart. In the first case, programming is generally effected by moving the industrial robot along the desired path by hand. In this case the path measuring systems are scanned during the time cycle and the measured co-ordinates stored point by point. The spatial proximity of these points to each other depends on the time cycle selected and on the speed of movement.

In the working cycle the stored co-ordinates are again generally read from the memory in the same cycle, and are predetermined as theoretical position values. The movement then takes place at the same speed as in the programming run. The speed, however, may also be varied by varying the time cycle.

Such a control system permits paths of any shape. With suitable compensation for the forces to be overcome by hand during the movement, it is very easy to program. However, it requires a large data memory (floppy disk, magnetic tape recorder). Moreover, the dynamic behaviour of the drives and controls is adapted to positions which must be approached in close temporal succession. Consequently, the positioning accuracy achieved in a point-to-point control system cannot be achieved here. Deviations occur in relation to the programmed path, particularly when the speed or load is increased in relation to the programming run, which causes sharp corners to be rounded, for example. In the second type of continuous path control system, only particularly important points on the path are stored. Not until the movement takes place is a closer succession of theoretical points produced by interpolation. At present only control systems with linear interpolation are available, but it is expected that the circular and parabolic interpolations normally applied in NC (numerical control) engineering (circular or parabolic arcs between the support points) will soon be offered.

2.3.2 Structure of the control system

Figure 2.15 shows the basic structure of an industrial robot control system. Its most important components are described in the following.

2.3.2.1 Information processing

The nucleus of the control system is the section described as information processing. This controls the program cycle, compares theoretical and actual positions,

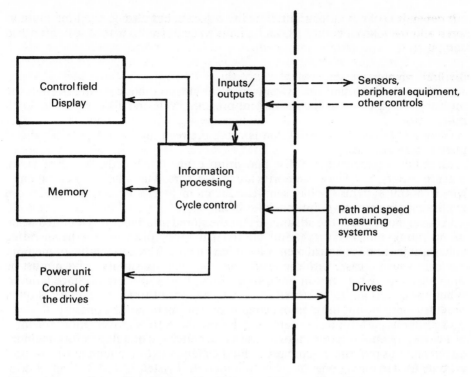

Figure 2.15   Structure of a control system for industrial robots.

controls the drives by means of the power section, and processes and stores the information fed in during programming.

In the first generation control systems, this section is constructed as a permanently wired circuit (relay or semi-conductor logic). The internal process control program is laid down by the selection and wiring of the components during manufacture. Subsequent modifications are always attended by the replacement or addition of components and corresponding wiring work. Today such control systems are only used for simple units of minimal flexibility. New control systems are equipped with a mini- or micro-computer as the central unit. The internal process control program is established essentially by computer programming, which from the very beginning makes for greater complexity, expressed in greater flexibility and much simplified operation, on the one hand, and on the other, modifications and extensions are much less troublesome as they can be carried out by reprogramming the computer.

Functions such as co-ordinate transformations, which allow, for example, the programming of articulated arm units in cylindrical co-ordinates, interpolation of synchronisation of the movement of the industrial robot with a moving assembly line or conveyor belt, are not conceivable without equipping the control system with a computer.

As flexibility increases still further, distribution of the tasks of the computer is expected to extend to several microprocessors, one of which provides the logical cycle control, and another, e.g. a special arithmetic processor, performs calculation tasks, whilst a further processor provides the supervision of the input and output operations and the operating elements. This could facilitate the combination of the control system of the industrial robot with other machines, e.g. machine tools, to form a flexible production cell, and simplify signal traffic with a superimposed computer.

2.3.2.2 Memory
A particularly important sub-system in any industrial robot control system is the memory, which contains the movement program cycle and also at least the path information.

Generally speaking, a distinction is made between mechanical, electrical and pneumatic memories.

Mechanical memories are disc and drum cams, plugboards, punched tapes, potentiometers, etc. These memories are in widespread use in older, permanently wired control systems. Today semi-conductor memories or core memories are mainly used in point-to-point control systems.

Core memories have the advantage that the stored information is retained when the operating voltage is switched off, or even in the event of a failure of the operating voltage, as in the mechanical memories. The contents of a semi-conductor memory are lost in such a case, with the result that the memory supply voltage must be supplemented by batteries in order to bridge at least short-term voltage failures. Where programs are to be stored for longer periods of time, the facility is often provided for recording the memory content on a magnetic tape recorder.

Continuous path control systems which store the path of movement in the form of a close sequence of points mostly use a magnetic tape or a floppy disk memory directly for this purpose. Pneumatic or fluidic memories (e.g. pneumatic flip-flops) are only used in purely pneumatic or fluidic control system (2.1, 2.2). Such control systems are mainly used for simple pick-and-place units. whilst mechanical memories also contain path information in analogue form (e.g. the position of a moving cam may identify the theoretical position value of an axis), data in electrical and pneumatic memories can only be fed in digital form and stored as such. Values measured by analogue means must therefore pass through an analogue-digital converter before storage.

2.3.2.2.1 Memory capacity
For point-to-point control systems and continuous path control systems with theoretical value generation by interpolation, the number of program steps that can be stored is the best indication of memory capacity.

However, in many control systems a program step only contains one, though in others several control functions (e.g. theoretical position and speed values, dwell times, link signals, etc.). Moreover, in some control systems one program step occupies a storage area which is always the same size, whilst in other cases its length depends on the information actually contained. Thus if the memory capacity is to be assessed on the basis of the number of program steps, the information and commands a program step contains must be known exactly.

In continuous path control systems in which a path of movement is stored as a close sequence of points, the memory capacity may be indicated by the number of storable points. However, because of the greater clarity it is more logical to use a maximum program cycle time for a typical scanning time cycle, e.g. 20 ms.

In addition to the memory capacity, the program capacity is also of interest, i.e. the number of main and subprograms which can be stored simultaneously and can be invoked independently. This capacity is only indirectly dependent on the size of the program memory and is determined essentially by the organisation of the memory management.

2.3.2.3 Path and speed measuring system

Path and speed measuring systems are not inherent components of the control system but they supply its input values.

The path measuring systems measure the path or distance covered during a particular movement or measure the instantaneous position of each axis of movement. The speed measuring system measures the corresponding instantaneous speed of each axis, also separately.

A distinction is made between analogue and digital path measuring systems. Analogue systems use mainly potentiometers, and with suitable wiring supply at their output a d.c. voltage which is proportional to the travel to be measured. Potentiometers are the cheapest, but also the least accurate solution. Digital systems measure either absolutely or incrementally. In both cases the distance to be measured is broken down into path quanta the length of which determines the resolution of the measuring system, and therefore limits the positioning accuracy of the unit.

Absolute path measuring systems (e.g. rotary encoders) supply full information at all times on the path to be measured. They operate with one or more coded discs which turn during the movement and are in most cases scanned photoelectrically. There is a unique relationship between the angle of rotation of the coded disc, and hence the path measured, and the derived binary coded numerical output value. Figure 2.16 shows the construction of such an encoder.

If a high resolution is required, the measuring systems are consequently large in size, and at the same time very expensive. Therefore incremental measuring systems are increasingly being used. Such measuring systems, for example, only count the path increments covered during the movement by photoelectric scanning of a rotating slotted disc. The counter reading then reproduces the instantaneous position of an axis. One disadvantage of this method is that the counter reading is lost when the unit is switched off or in the event of a voltage failure. The industrial robot must then be brought to a reference position in which all the counters are zeroed or set to a defined value.

A further disadvantage is the propagation of a single counting error to the measurement of all succeeding positions. In addition to photoelectric systems so-called resolvers are in widespread use. These operate electromagnetically and supply a sinusoidal voltage at their output. For rough measurement, each reversal of this voltage is counted, whilst accurate measurement is carried out between two voltage reversals by phase comparison with a reference voltage.

The speed of movement is measured either direct, by means of special tacho-generators, or indirect by suitable evaluation of the signals of the path mesuring system.

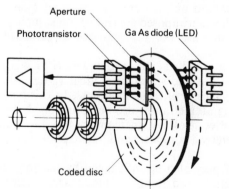

Aperture

Phototransistor   Ga As diode (LED)

Coded disc

Figure 2.16
Construction of a rotary encoder.

The drive of industrial robots by means of stepping motors represents a special case.

Since these motors can be controlled to take up accurately a desired position by providing in advance a certain number of path increments, path measuring systems are superfluous in this case.

One disadvantage is that such motors may lose steps at high loads or after collisions with obstacles, i.e. may perform fewer steps than predetermined in the control system. Thus the positioning will be incorrect for all succeeding positions.

Simple pick-and-place units, which are generally driven pneumatically and positioned by means of mechanical end stops, do not of course require any path measuring systems.

2.3.2.4 Interfaces with peripheral units

Since an industrial robot does not generally operate independently of other machines, but is integrated in a production process, flow must be made for signal and data flow and transmission to the surrounding area. In the simplest case, this information exchange runs through so-called input and output ports. For example, a signal which is transmitted by the industrial robot control system after a certain program step has been executed, through one of its output ports, could inform a machine tool control system and that the insertion of a workpiece by the industrial robot has been completed and that the machining process may commence.

On the other hand, the industrial robot must then wait to unload the machine until the machining process has been completed. It may therefore be programmed so that it does not commence executing a particular program step until a signal is applied to one of its input ports from outside.

Some control systems have additional inputs for so-called positioning switches, which are used to perform a search movement. The gripper is moved along a path on which a workpiece is expected, e.g. from above onto a workpiece stack. As soon as the positioning switch, perhaps an inductive or optical proximity switch, grips the workpiece, the search movement is interrupted and the uppermost workpiece can be gripped.

Some control systems permit the connection of external tachogenerators or torque pick-ups, in order to synchronise the movement of the robot with that of a moving conveyor belt, for example.

PPI S-20, Cincinnati

For more comprehensive data flow, such as that required for the example, where several industrial robots are to be controlled by a computer or where complex sensors are to be connected, these inputs and outputs are no longer adequate. Uniform interfaces for the data traffic with superimposed control systems, such as those known in data transmission systems, are not yet available as standard equipment.

However, some control systems permit a so-called external data feed (PPI S-20) by means of special interfaces. Both position and speed theoretical values and sub-program addresses may be fed in by means of these interfaces. In this way positions which are not permanently programmed may also be approached, i.e. their co-ordinates are not located in the control system memory but are first measured by external sensors or calculated by a computer and then transferred to the control system.

Furthermore, the program cycle itself may also be influenced by feeding in conditions from outside for program jumps and jump addresses. Orientation

22

devices in which an industrial robot is linked by these interfaces to a television sensor have already been demonstrated by the Fraunhofer Institutes for Production Engineering and Automation, Stuttgart, and for Information Processing in Engineering and Biology, Karlsruhe. The sensor measures the position and orientation of randomly presented workpieces, converts the measured results into axial co-ordinates of the industrial robot and transmits them to the control system for gripper positioning (2.5).

2.3.3 Programming

2.3.3.1 Program structure

The movement program of an industrial robot is a sequence of instructions which are executed according to internal specifications laid down in the control system. In the case of point-to-point controlled units, the program consists of individual program steps, each program step generally containing all the instructions and information required for the movement from one position to be approached to the next, i.e. at least procedural commands for all axes to be travelled, and in addition theoretical position values or data on the memory containing the theoretical position values, in the case of freely positionable units. Furthermore, a program step may contain desired speed values, dwell times, input and output commands, jump commands and further additional functions. In most individual applications, all the program steps are executed in succession and always in the same sequence. However, modern control systems also provide the possibility of branching in the program cycle, to perform even more complex tasks such as the interaction of an industrial robot with several machines with different cycle times. In such cases the division of the movement program into a main program and several sub-programs facilitates programming and increases the flexibility of the unit considerably, particularly when the sub-programs may be called up externally (controlled by the process).

2.3.3.2 Programming methods
Industrial robots, by contrast to machine tools, are almost always programmed at their work place.

In some robots the sequence of movements and the position co-ordinates are programmed separately, and the processing program can then often be established independently of the industrial robot, e.g. by punching tapes. However, external programming of the position co-ordinates would be extremely complicated, since the geometrical relationships between industrial robot, workpiece and machines to be operated, for example, are not generally known sufficiently accurately, and would, moreover, have to be converted to the co-ordinate system of the robot axes, which is often extremely involved.

When programming at the work place (direct programming), the program is always established by means of the industrial robot itself. The machine is therefore unusable elsewhere in the programming phase. Two programming processes are distinguished, but they are almost always mixed together.

2.3.3.2.1 Teach-in
In the teach-in programming process, the movement information is recorded wholly or partially during a programming run in which the movement to be executed is fed to the machine. In the case of point-to-point controlled industrial

Figure 2.17:
Manual programming unit by means of
which a point-to-point controlled
industrial robot can be operated by
hand.

robots the machine is moved from one position to the next by means of special operating or control elements. At each point the position co-ordinates are stored as theoretical values or set points by pressing keys. In most units, other information, such as desired speeds, input/output signals, dwell times, may also be fed in. The control elements required for programming are often installed on a portable manual programming unit, so that the programmer is able to observe the positioning directly on the gripper or tool.

Information which cannot be recorded during the programming run is then fed in by means of a keyboard or similar device.

Using the manual programming unit illustrated in figure 2.17 any axis of the industrial robot can be individually controlled.

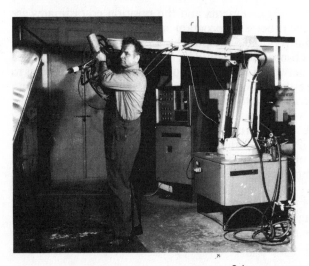

Figure 2.18:
Programming of a
continuous path
controlled industrial robot
for painting.

24

Figure 2.19:
Keyboard of an industrial robot
control system. The data fed in are
displayed on the screen of a
visual display unit.

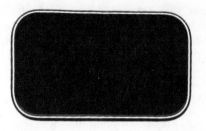

Articulated arm units with less clearly defined axis configurations are in most cases programmed in a simpler system of co-ordinates, e.g. cylindrical or Cartesian co-ordinates (ASEA). To obtain a movement along one axis of such a system of co-ordinates several industrial robot axes must generally be moved at the same time. The co-ordinate transformations required for this are carried out by the computer inside the control system. In the case of the 6 CH-Arm industrial robot from Cincinnati, which is equipped exclusively with rotary axes, the position of the gripper or tool is programmed in Cartesian co-ordinates and the orientation in spherical co-ordinates.

Different gripper or tool sizes may be fed into the control system so that the programmed position, and hence the origin, of the co-ordinate system, is always the actual operating point.

Continuous path controlled industrial robots, whose control system stores the path of movement by means of a close succession of individual points, are in most cases also programmed by teach-in, i.e. by going through the work cycle. The movement here, however, is effected not by control of the drives but by direct manual guidance of the axes. For example, if a painting operation is to be programmed, the programmer sprays a sample workpiece, using the spray-gun secured to the industrial robot, and in doing so also moves the entire industrial robot. During the movement the co-ordinates of all the axes are scanned in a fixed time cycle, are stored, and are then available for the work cycle as the desired position. Simple additional functions, such as "Spray-gun on", "Spray-gun off", are also recorded during the programming run (Figure 2.18).

One problem posed in this programming process is the kinematically dependent forces which the programmer must overcome. Although measures are taken with all equipment to reduce these forces, they are in most cases relatively high, particularly at the boundaries of the work area.

For this reason, a very light frame, identical in kinematic structure, was developed for an industrial robot now on the market, which is guided during programming instead of the actual robot, and which therefore facilitates the work of the programmer.

2.3.3.2.2 Manual data input

The term manual data input embodies all processes in which the program or parts of it are manually fed into the program memory of the control system by means of keyboards, decade switches, by setting mechanical memories, etc. This process is used exclusively in older, simple control systems, but in newer control systems, particularly computer control systems, only for correcting and supplementing the program established by teach-in. In the latter case alphanumerical keyboards (Figure 2.19) are mainly used to read out, read in or modify all the program elements and data. A case more frequently encountered is the programming of program sequence and position co-ordinates and some speed steps, for instance, by teach-in, by means of a manual programming unit. Dwell times, link signals, program jumps, program limits, etc. are then programmed by means of a keyboard.

In older control systems, which are programmed exclusively by manual data input, there are a large number of variants, only some of which may be called typical. The program cycle may be established by changing over electrical leads or pneumatic tubes, by setting or modifying diode matrices or bus bar distributors, by setting cams or by punching tapes or cards.

To store the desired position points, mechanical memories, which can be set manually, are used. In simple systems, these are mainly adjustable stops, and in units with a continuous path measuring system, potentiometers.

2.3.3.2.3 Programming languages

There are several methods of facilitating the programming of industrial robots by means of programming languages. As far as industrial applications are concerned, only the relatively simple programming language SIGLA, developed by Olivetti for the SIGMA industrial robot, has so far been used.

The language is on the level of assembler languages, i.e. every function to be executed by the robot is individually formulated in the programming language and fed in via a computer terminal. The instructions for this language include, for example, conditional jumps, setting and inquiring from counters, two-arm co-ordinates and evaluation of signals from a load cell. However, the position programming may also be carried out in the teach-in mode in this case.

Higher, problem-oriented programming languages, which are intended not only to be used for programming isolated industrial robots, but also more complex systems, such as flexible assembly or production cells, are being developed at several centres but have not yet got beyond laboratory studies (2.6, 2.7, 2.8).

2.4 Grippers

2.4.1 Introduction

Grippers for industrial robots have the task of making the connection between workpiece and robot. According to the application, however, grippers must also perform more advanced functions. For example, they must centre and orientate workpieces, perform additional movements or feed information on workpieces to

Figure 2.20
Special gripper for pipes.
(Photograph: Robert Bosch
GmbH).

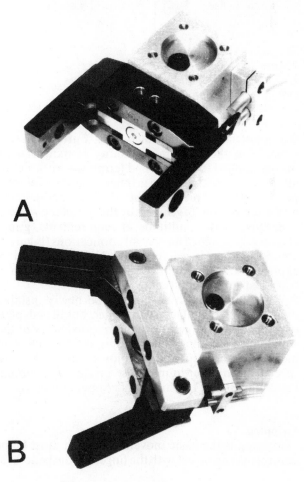

Figure 2.21a
Gripper components
A – Parallel finger
movement
B – Rotational finger
movement
(Photograph: Fibro).

A

B

27

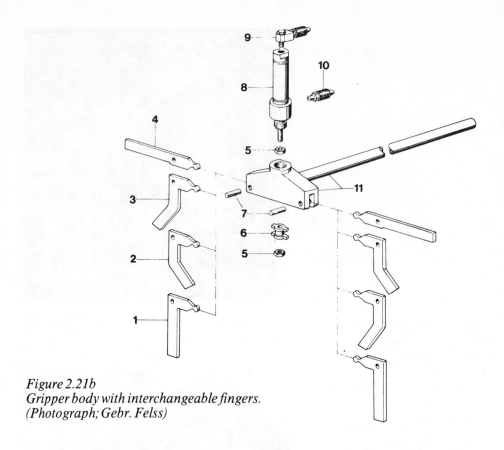

Figure 2.21b
Gripper body with interchangeable fingers.
(Photograph; Gebr. Felss)

the industrial robot control system. Furthermore, grippers should be as light as possible, as the acceleration ability of industrial robots depends greatly on the masses applied to the arm end (gripper + workpiece), and therefore influences the attainable cycle time. The gripper weight therefore also influences the maximum permissible workpiece weight. Moreover, close limits are imposed in many cases on the dimensions of grippers by the size of the working area available.

Because of the multitude of requirements, grippers have so far been almost exclusively designed for special applications (Figure 2.20). In all cases basic gripper components are used (Figure 2.21) which must be adapted to the particular application, and the aim has been to provide them with the capability to grip the widest range of workpieces.

Where industrial robots are to be newly installed, the grippers are supplied mostly by the manufacturer or seller, but the adaptation and further development of grippers is then generally the responsibility of the user. The special problems which arise in connection with grippers therefore make it necessary for users of industrial robots to train employees from their own firms to become gripper specialists, in order to minimise production, commissioning and down times for handling systems. Definitions of gripper functions are given below:

Gripping
"Gripping is the basic movement which must be performed to obtain sufficient control over an object with the fingers or the hand". (2.9).

28

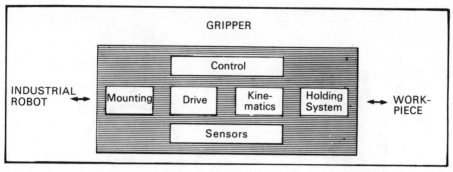

Figure 2.22   Gripper with its sub-systems.

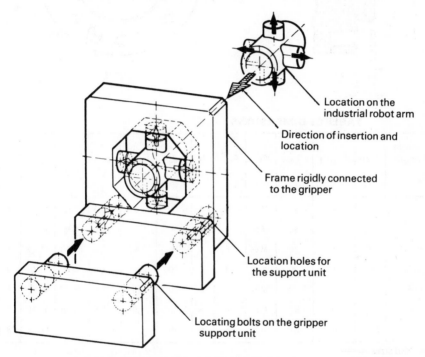

Location on the
industrial robot arm

Direction of insertion and
location

Frame rigidly connected
to the gripper

Location holes for
the support unit

Locating bolts on the gripper
support unit

*Figure 2.23   Diagrammatic representation of an interchangeable gripper
mounting plate. (Photograph: VW AG).*

Clamping
"Clamping is the retaining of the workpiece in a given position". (2.10).

Releasing
"Releasing is the opposite of clamping". (2.10).

The definitions given are not specific for industrial robot grippers and an attempt is therefore made in the following to give suitable definitions.

Industrial robot gripper
"The gripper is the sub-system of an industrial robot which effects the transmission of load from the workpiece to the industrial robot in order to secure the position of the workpiece in relation to the robot".

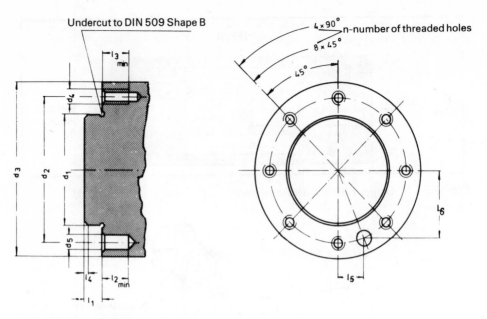

Undercut to DIN 509 Shape B

$4 \times 90°$ n-number of threaded holes
$8 \times 45°$

TABLE OF DIMENSIONS

Diameter Length Size	d_1 f7	$d_2 \pm 0,1$	d_{3min}	d_4 [1]	d_5 H7	l_1	l_{2min}	l_{3min}	$l_{4min} \times 45°$	$l_5 \pm 0,02$	$l_6 \pm 0,02$	n
1 - - - - -	32	40	50	M 6	6	4	8	8	0,8	7,65	18,45	4
2 ————	40	50	60	M 6	6	4	8	8	1,0	9,55	24,00	4
3 - - - - -	50	60	70	M 6	6	4	8	8	1,0	11,50	27,70	4
4 ————	63	75	90	M 8	8	6	10	10	1,0	14,35	25,40	4
5 - - - - -	80	90	105	M 8	8	6	10	10	1,0	40,20	41,55	4
6 ————	100	110	125	M 8	8	6	10	10	1,6	21,05	50,80	8
7 - - - - -	125	135	150	M 8	8	6	10	10	1,6	25,85	62,35	8

Standard sizes ————
Intermediate sizes – – – –

(1)   In exceptional cases through holes are permitted for the threads.

Figure 2.24   Standard proposed by the Arbeitsgemeinschaft Handhabungssysteme (Gripper Working Group) for design of the mechanical interface on the industrial robot for attaching the gripper.

2.4.2   Construction of grippers

Grippers consist of different sub-systems (Figure 2.22), which are discussed in sections 2.4.2.1 to 2.4.2.5.

Grippers need not consist of all the elements such as mounting plate, drive, kinematics, retaining system, sensors and control system (Figure 2.22). A gripper

30

may in the simplest case incorporate only the mounting plate and the holding system. (A magnet or hook at the end of the robot arm). Non-powered grippers are in widespread use in conveyor engineering. Normally, however, grippers are used which incorporate all the sub-systems.

2.4.2.1 Mounting plate

The function of the mounting plate is to make a secure connection between the industrial robot arm and the gripper. If several grippers are to be used for handling different workpieces with an industrial robot, the mounting must allow the grippers to be interchanged.

The gripper change is generally carried out manually, but automatic changing mechanisms have already been proposed (e.g. by ASEA, Renault, IPA-Stuttgart, Olivetti, Unimate, Kawasaki and VW (Figure 2.23), which have a suitably designed interchangeable mounting plate (Section 2.4.4.2: concepts for flexible grippers).

A standards proposal, which was developed as part of the HDA programme promoted by the BMFT (2.11) by the Arbeitsgemeinschaft Handhabungssysteme (Gripper Working Group), is presented in Figure 2.24.

2.4.2.2 Drive

Gripper drives have the task of converting the power supplied to the grippers into kinetic energy used for moving the linkages (Figure 2.25). The function which supplies the vacuum for suction devices is not termed a gripper drive since the low pressure is not converted to movement but is used to apply the retaining forces.

Figure 2.25
Gripper drive
function.

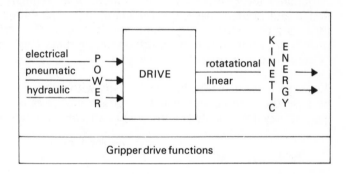

Gripper drive functions

Figure 2.26
Gripper drive
types.

Gripper drive	Drive movement
1 Electrical drive	
1.1 Stepping motor	rotational
1.2 D.c. motor	rotational
2 Pneumatic drive	
2.1 Pneumatic cylinder	linear
2.2 Compressed air motor (high speed)	rotational
2.3 Swivel cylinder (low speed, angle of rotation limited)	rotational
3 Hydraulic drive	
3.1 Hydraulic cylinder	linear
3.2 Hydraulic motor (angle of rotation unlimited)	rotational
3.3 Swivel cylinder (angle of rotation limited)	rotational

31

The gripper drive types shown in Figure 2.26 are now in general use.

The commonest drives are pneumatic cylinders, since they are cheap and light. Moreover, the energy supply to the cylinders with flexible hoses causes the least difficulties, in most cases, compared with hydraulic pipes or cables for electric motors. Because of the small overall size, hydraulic cylinders are often chosen for large gripping forces. The drive power for industrial robots should, if possible, also be used for the gripper drive, to simplify the power supply system.

2.4.2.3 Mechanics

There are a large number of mechanisms available for converting the linear or rotational drive motion to a linear, rotational or curvilinear finger movement. For the sake of clarification, these are divided as follows:

1. Speed ratio. The motions of industrial robot grippers may exhibit the following speed ratio variation (ratio of the drive speed to the speed of the fingers):
- speed ratio constant   over the finger stroke
- speed ratio increasing   over the finger stroke
- speed ratio decreasing   over the finger stroke
- speed ratio with minimum over the finger stroke
- speed ratio with maximum over the finger stroke

For grippers it is advisable in most cases to use mechanisms which operate either at a constant speed ratio or at a ratio which varies during the finger stroke in such a way that as the clamping diameter increases, so do the clamping forces applied, since the workpiece weight generally increases with the clamping diameter. Articulated lever mechanisms may indeed produce high speed ratios with simple design and limited space, but the clamping range within which large clamping forces are produced is very small, with the result that articulated lever grippers must be set accurately to a certain operating range (Figure 2.27). See also 2.4.4.3 "Calculation of grippers".

2. Finger movement. Simple mechanisms with rotational movement are often used because they are economical to produce and because of their uncomplicated design, they are reliable (Figure 2.28) particularly when pneumatic cylinders are

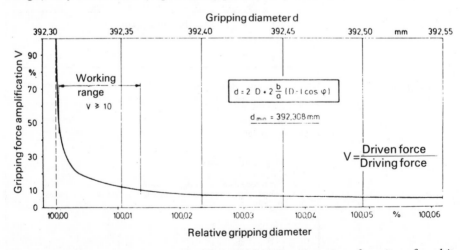

Figure 2.27   *Gripping force of an articulated lever gripper as a function of working diameter (according to (2.12)).*

32

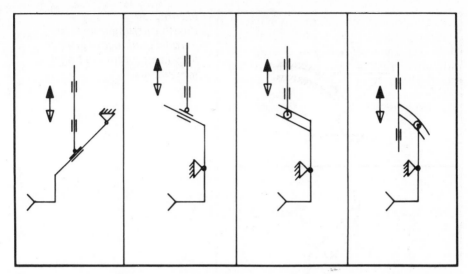

*Figure 2.28   Possible arrangements for converting a linear drive movement
(cylinder) into a rotational finger movement.*

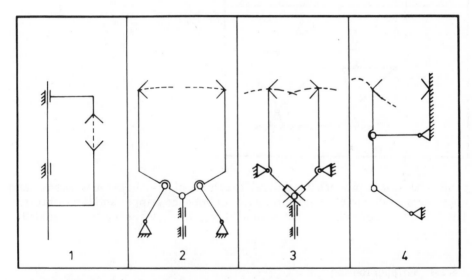

*Figure 2.29   Mechanics of grippers with different finger movements: 1. linear
movements; 2. approximately straight-line finger movements; 3.
rotational finger movements; 4. curvilinear finger movements.*

used as the drive system. However, one of their disadvantages is that they require
more programming expenditure than mechanisms which provide linear finger
movements (Figure 2.29). This is because the finger movement is simpler to assess
by the programmer in the latter case and is therefore less prone to cause collisions.

In order to grip workpieces with parallel faces, it is recommended that the gripper
fingers also be guided in parallel (Figure 2.30), to ensure a large area support of the
fingers where there are tolerance variations or differences in the distance between
the faces. However, it is not advisable in most cases to guide both gripper fingers in

33

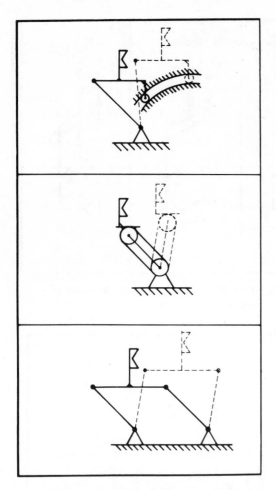

Figure 2.30
Possible arrangements for
parallel guidance of gripper
fingers for rotational finger
movements.

parallel because in practice it is difficult to achieve alignment of the workpiece and finger surfaces due to elastic deformations of both the gripper and the workpiece. The mounting of one of the fingers should be compliant, while the other is guided in parallel.

2.4.2.4 Gripper tooling
The function of the tooling is to fix the position of the workpiece in relation to the gripper kinematics and ultimately in relation to the industrial robot. The tooling must have shaped elements for this purpose, which correspond to the shape of the workpieces. The workpiece may be held by clamping or by its shape or as a result of the properties of the tooling material, i.e. by normal, frictional or magnetic forces.

Figure 2.31 shows possible surface pressure systems (normal forces) which are also required for producing frictional forces. Also pressure may be provided, for example, by adhesive materials.

2.4.2.5 Control system and sensors
Reference is made to the chapter on control system and sensors in this book with regard to the control system and sensors which are active in the gripper system.

Figure 2.31
Possible methods of
producing surface
pressures.

> Means of holding workpieces
>
> – gravity
>
> – mechanical clamping between gripper fingers
>
> – vacuum suction, vacuum pumps and Venturi suction devices
>
> – magnetic forces (permanent magnet, electro-magnet)

2.4.3 Gripper performance specification
General specifications and application-specific specifications may be issued for the design of the grippers.

2.4.3.1 General requirements
The general requirements for grippers are reproduced here in the form of a table, in which completely contradictory requirements are set out side by side. An evaluation from which priorities are made cannot be undertaken until a special performance specification has been established.

Grippers should:
be as light as possible in order to minimise the static and dynamic load of the handling systems (Figure 2.32);
be as small as possible in order to reduce the space required for the grippers in the work area;
be as rigid as possible, so that the workpiece position can be retained as accurately as possible;
act on the workpieces with as much gripping force as is required and with as little as possible, to avoid damage;
be as reliable as possible;
be able to be designed and manufactured with the least possible expenditure;
require as little maintenance expenditure as possible.

Figure 2.32
Example of a
lightweight gripper.
(Photograph: VW
AG).

2.4.3.2 Special requirements

The information given here is required for designing grippers. The boundary conditions or performance with which the gripper must comply must necessarily be determined from this information. Under certain conditions a largely comprehensive gripper performance specification has to be prepared for a particular workplace requiring an effort that should not be underestimated.

Table 1: Workpiece data:
- dimensions, tolerances
- weight, position of centre of gravity
- material
- magnetic properties
- surface, permissible surface pressure, roughness, contamination (chips, dust, oil, water, etc.).
- permissible stresses
- temperature

The information on "dimensions", "weight" and "material" and "magnetic properties" may be on the basis of *workpiece drawings and parts lists.* Measures must be taken to ensure that the drawings relate to the condition of the workpieces during the handling process. The information on "surface condition" and "permissible stress" can in most cases be obtained by *visual inspection* thus not only the drawings, but also the workpieces themselves, should be presented in the handling condition. The "temperature" information is obtained from the production cycle.

The terms 'donor'* and 'receptor'* used in the following should be explained here. In workpiece handling, the workpieces are selected before handling, and this device (e.g. chuck) should be termed the escapement. The workpieces are then transferred by the industrial robot and deposited into the feeder device (e.g. magazine).

Table 2:
Information on donor* and receptor*
1. Position and orientation of the workpieces in the donor (with tolerance data) and in the receptor.
2. Clearances around the workpieces in the donor and receptor.
3. Position and dimensions of the workspace for the industrial robot and the workpiece in the donor and receptor.
4. Permissible and required forces and moments for mating in the donor and receptor.

Table 3:
Industrial robot data
1. Available gripper drives.
2. Available control information (inputs, outputs).
3. Loading capacity of the industrial robot.
4. Acceleration of the industrial robot.
5. Mounting plate dimensions.
6. Kinematics of the industrial robot.
7. Working area of the industrial robot.

---

*Translator's note: The original text uses the acronyms ENE (Entnahmeeinrichtung) and EGE (Eingabeeinrichtung).

36

Table 4:
Standards and guidelines
– delivery instructions of the user
– predetermined design shapes for grippers or components of grippers
– permissible costs
– safety allowances
– emergency stop performance

Table 5:
Sequence of events
– cycle diagram
– permissible closing time
– permissible opening time
– permissible changeover time

If the information listed in Table 1 to 5 is provided, work can be started on the development of a detailed gripper performance specification, and the following requirements for the gripper, among other things, can be determined (Table 6). Further requirements, e.g. with regard to the permissible closing time of the gripper or permissible costs, may be obtained directly from the information mentioned in Tables 1-5.

Table 6:
Gripper requirements
1 – Permissible gripper weight (W perm.)
2 – Workpiece gripping areas and surfaces
3 – Opening (capacity) of the gripper
4 – Finger movement
5 – Closing forces

With reference to 1. the permissible gripper weight, which is determined according to the following rough estimate, presupposes that the moment exerted by the gripper and the workpiece on the industrial robot arm may be ignored. Thus the gripper and workpiece must not be overloaded. Information is given on the permissible moment on the gripper mounting by the industrial robot manufacturer.

W perm. = Safe load capacity – maximum workpiece weight

With reference to 2. the possible gripping areas are established on the basis of an evaluation of the information on the position of the workpieces in the unloading or loading device. In establishing the gripping areas, consideration must be given to the range of gripper insertion movements, with its tolerances (e.g. due to vibrations) and to the area of movement of the gripper elements during opening and closing.

The possible gripping areas may be obtained from the available workpiece drawings, and a separate indication (e.g. identified by different colours) for the unloading and loading devices would be helpful.

It must then be ascertained whether there is correspondence between the possible gripping areas in the donor and receptor devices.

Case I: If there is *no correspondence,* the following solutions are possible:
    1. Intermediate depositing of the workpiece between escapement and feeder for re-gripping.

37

2. Modification to the escapement or feeder devices.
3. Modification to the workpiece.

In most cases a solution according to 1. is not possible because of increased cycle times and cost. Workpiece modification moreover, may only be carried out in rare cases.

The remaining solution is 2.

In this connection reference is also made to the importance of suitable gripper adapted workpiece designs.

Case II: If there is only *correspondence* in relation to *one workpiece surface,* which does not permit clamping between between two opposing faces, it must be decided whether a suction or magnetic gripper may be used. If none of these can be used, one of the three possible solutions of Case I must be used.

Case III: If there is *correspondence of several individual gripping surfaces* without pair formation, the same procedure as for II must be adopted, or a three- or multipoint finger grip must be provided.

Case IV: If *pairs of gripping surfaces* are produced which permit clamping between two fingers, the most favourable pair must be determined. In this case the whole spectrum of workpieces should be considered to establish correspondence of pairs of gripping surfaces within a wider range.

Following the analysis of the gripping surface or area information must be available on the distance between the gripping surfaces and their shape.

The *opening* of the *gripper* (requirement 3 in Table 6) may be determined from the distance between gripping surfaces, with the following possibilities:

1. The opening of the gripper is designed so that it bridges the entire distance between the gripping surfaces within the workpiece spectrum.

2. The opening of the gripper is designed so that the maximum *tolerance* of a distance between gripping surfaces is bridged. The entire spectrum of the gripping surface distance is then covered by changing the fingers (Figure 2.33).

On finger movement, for most applications the simple design of the grippers with rotary finger movement will be sufficient. Articulated kinematics which provide approximate or accurate parallel guidance, e.g. by means of non-friction bearings, have so far only been used under special application conditions, e.g. for research purposes or for manipulators in nuclear power stations.

Figure 2.33
Stepped capacity adjustment on a double gripper. (Photograph: Böhringer).

On closing forces the greater the extent of clamping, the smaller are generally the closing forces required (2.4.4.1). The calculation of the holding forces is shown in 2.4.4.3.

2.4.4 Designing grippers

A universally applicable route from the performance specification to the design of an optimum gripper cannot be developed within the context of this contribution. However, particularly important problems and their possible solutions are discussed in detail. Reference is also made to Section 2.4.2 "Gripper construction" and 2.4.5 "Gripper manufacturers", which under certain circumstances may provide suggestions for gripper design.

2.4.4.1 Positioning of workpieces in the gripper

On the basis of the gripping surfaces established in the performance specification, gripper fingers may be developed which ensure sufficiently secure clamping according to the forces applied to the workpiece.

Among the possibilities for fixing the position of bodies mentioned in Section 2.4.2.4 "Gripper tooling", particular consideration will be given here to pressure grip and form-dependent gripping, and the associated means of producing a pressure grip.

In form-dependent gripping, the gripping forces are transmitted as normal forces between the active surfaces. On the other hand, in pressure gripping the gripping forces are transmitted as transverse forces between the active surfaces by friction. The transmitted forces depend in this case on the gripping force and the static friction coefficient between the workpiece surfaces.

By contrast to form-dependent gripping, normal forces must be applied, before positional fixing operates. If these normal forces for producing a frictional force are not present, e.g. due to energy failure, the position fixing is no longer guaranteed. It is therefore important to fix the position of workpieces in relation to the industrial robot as far as possible by form-dependent gripping. Procedures for achieving as large a proportion of this type of gripping as possible are described in the following.

Various possibilities result for producing form-dependent gripping, depending on the gripping surfaces.

Cylindrical and prismatic workpieces

If cylindrical workpieces are gripped on the cylindrical surface, form-dependent gripping and centring may be achieved by means of two, three or more point or line systems (Figure 2.34).

A three-point system is achieved, for example, by means of a prism with back stop. The advantage of this system over a system with two or more points is that with the three-point system positive centring is obtained. An example of a completed gripper with centring is shown in Figure 2.35.

Centring problems are encountered especially with workpieces having wide tolerances. A solution to this problem is shown in Figure 2.36, another solution in Figure 2.37.

Two-point centring   Three-point centring

 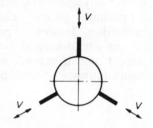

Four-point centring   Three-point centring

Figure 2.34   Possible methods centring cylindrical workpieces.
  V = closing speed,
  δ = prism opening angle,

$$f(\delta) = prism\,factor = \frac{1}{\sin \delta/2}$$

Irregular workpiece shapes
In the case of irregular workpiece shapes, form-dependent gripping may be achieved by a number of different procedures.

Procedure A:
Production of a negative shape for the workpieces by cutting or forming the gripper fingers (Figure 2.38 and 2.39).

Figure 2.35   Gripper with four-point centring (Photograph: Kuka (2.13)).

*Figure 2.36   Operating sequence and gripping position for forging a rear-axle
shaft flange (Photograph: Kuka (2.13)).*

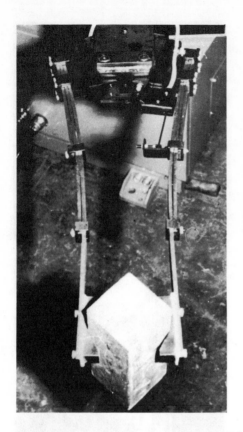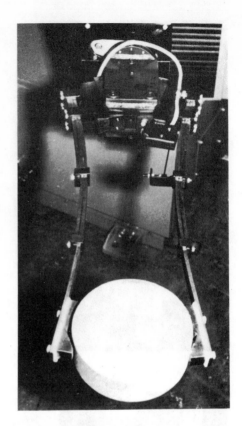

Figure 2.37   Flexible gripper fingers (Photograph: Kuka (2.13)).

*Figure 2.38
Form-dependent
gripper
(Photograph: VW
AG)*

*Figure 2.39
Form-dependent
gripper for body
parts.
(Photograph: VW
AG)*

Procedure B:

Production of a negative shape for the workpieces by moulding with plastics
(Figure 2.40).

By moulding the workpiece or special shapes with plastics (2.14) form-
dependent gripping is obtained which gives secure, positioned gripping of
workpieces (Figures 2.41, 2.42).

*Figure 2.40
Construction of a
moulding finger with
reinforcing pins (2.14).*

Key:
1. Support for finger insert
2. Finger insert
3. Workpiece
4a and 4b Reinforcing pins
5. Additional support
6. Gripper finger

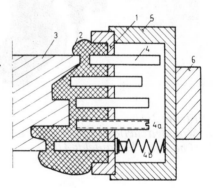

Figure 2/41
Moulded suction gripper for a pipe bend.
(Photograph: Kaufeldt AB).

without workpieces

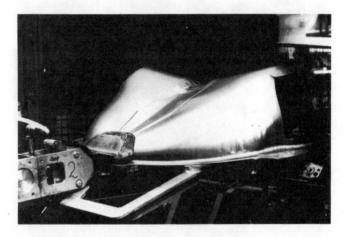

Figure 2.42   Moulded fingers with articulated lever gripper for body part (Photograph: VW AG).

45

Procedure C:
Production of a negative shape for the workpieces by moulding with interchangeable gripper fingers, e.g. with a stack of laminations which are clamped after manual moulding (Figure 2.43).

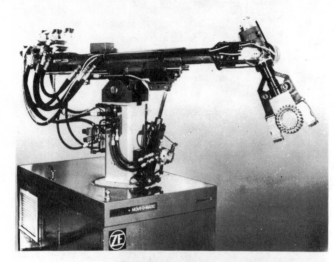

Figure 2.43 Laminated gripper (Photograph: ZF AG).

Figure 2.44 Example of a diaphragm gripper. (Photograph: Carl Freudenberg).

Procedure D:
Production of the negative shape by means of elastic gripper fingers or diaphragms. In most new installations of industrial robots, modification of the grippers is required. It is therefore recommended that these grippers be made interchangeable, particularly in view of the wear on the contact faces of the gripper fingers (2.51).

2.4.4.2 Designs for flexible grippers

The special problems which arise if several different workpieces are to be handled by one industrial robot are discussed in this chapter.

If grippers are to be used in small and medium series production, the number of requirements for the gripper increases in comparison with the use of the gripper in large series production, since there are generally several workpiece shapes, the workpiece weights vary and the inlet channels of the donor and receptor device have different designs. Examples of the possible applications of flexible grippers are shown in Figure 2.45, whilst Figure 2.46 shows the effects of increased flexibility. In the case of workpieces handling in small and medium series assembly processes an even larger number of different workpieces must be gripped. It is therefore understandable that since the flexible handling systems are developed for small and medium series attention is now being directed towards designing grippers so that they are as flexible as other parts in the handling system.

The problems which arise have led to a situation where grippers have not yet achieved the universality of other sub-systems (e.g. the control system) of the handling systems on which they are used. Problems in the design of grippers and also with the peripheral equipment of industrial robots, particularly the orientation devices, often lead to a situation where economic use of handling systems in small and medium series production is not possible. Factors which must be considered when constructing grippers are listed in Figure 2.47 (factors which have a particularly marked influence on the flexibility of the grippers are given special mention).

The gripper objects, with their charateristics of shape, size and weight, have a decisive influence on flexibility. The effect of the work space in which the gripper is to be used relates in particular to the size of the gripper, which must be dimensioned on the basis of the smallest occurring work space for the application, and therefore represents a size limitation for the gripper. The loading capacity of the industrial robot limits the permissible weight of a gripper. (Figure 2.48).

It is generally the case that the gripper for a particular workpiece spectrum is more permissible in its use, the lighter and smaller it is. With a smaller gripper, the permissible workpiece volume will increase, and with a lighter gripper the permissible workpiece weight will increase.

Consequently extremely important features of gripper flexibility are clamping range and capacity, adaptability of the gripping surfaces, gripper weight and gripper volume. Possible solutions have been put forward by Auer (2.12), for example, to the problems associated with increasing clamping ranges and clamping force. The methods of solving problems of shape adaptation are discussed in the following.

Grippers adapted to workpiece spectrum

The weight distribution of workpieces for handling operations to be performed by industrial robots is given in Figure 2.49.

The weights vary from below 50 g to over 100 kg. It is illogical to construct a gripper which is capable of covering the entire range of gripper weights, since the size and weight of such a gripper would greatly limit its possible applications in the handling of small parts in particular. The whole spectrum of workpiece weights should therefore be divided into sub-ranges which are covered by different flexible grippers.

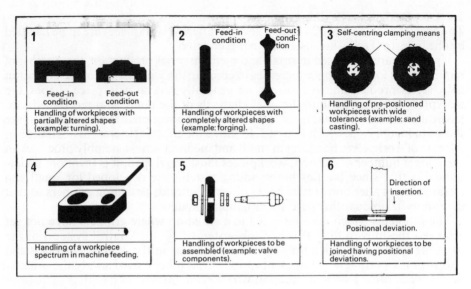

Figure 2.45   Possible applications for flexible grippers.

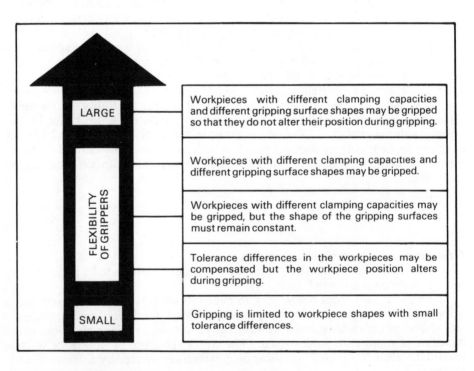

Figure 2.46   Effects of increased gripper flexibility.

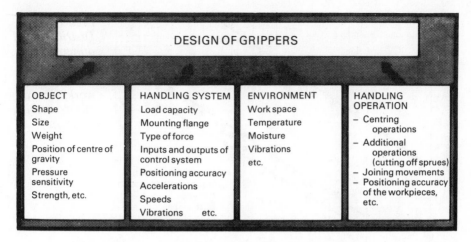

DESIGN OF GRIPPERS

OBJECT	HANDLING SYSTEM	ENVIRONMENT	HANDLING OPERATION
Shape	Load capacity	Work space	– Centring operations
Size	Mounting flange	Temperature	– Additional operations (cutting off sprues)
Weight	Type of force	Moisture	– Joining movements
Position of centre of gravity	Inputs and outputs of control system	Vibrations	– Positioning accuracy of the workpieces, etc.
Pressure sensitivity	Positioning accuracy	etc.	
Strength, etc.	Accelerations		
	Speeds		
	Vibrations etc.		

Figure 2.47   Factors influencing the design of grippers.

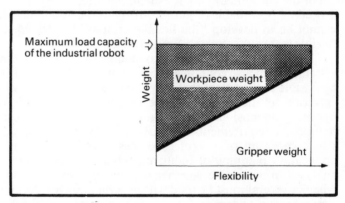

Figure 2.48   Relationship between the load capacity of an industrial robot and the weight of the workpiece and gripper.

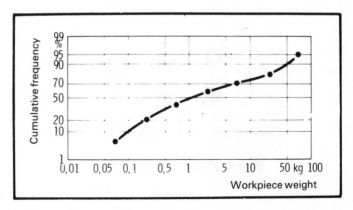

Figure 2.49   Distribution of workpiece weights for handling operations.

49

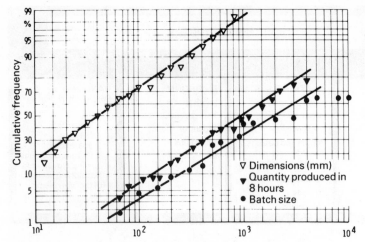

Figure 2.50
Distribution of
dimensions,
quantities and
batch sizes of
workpieces at
105 assembly
locations.

For the consolidation of the clamping range and capacity of grippers, a similar procedure is available (Figure 2.50). Thus the aim of developing flexible grippers cannot be to develop "the flexible gripper" but to develop grippers which are flexible within a reasonably limited range.

There are on the market various module systems for the construction of handling systems which also contain gripper components. These gripper components must be flexible to the extent that they can be adapted to the widest possible workpiece spectrum by the addition of special gripper fingers. However, grippers are not optimally dimensioned with regard to weight and volume until they are developed to meet the requirements of a workpiece spectrum and are then able to cover the whole spectrum of workpiece shapes and sizes without being changed. In the application planning of industrial robots, the available workpiece spectrum is analysed to determine whether it is possible to use a single flexible gripper, or whether several must be used within certain ranges. As an example we show here the procedure adopted for developing a special flexible gripper for a given workpiece spectrum (Figure 2.51).

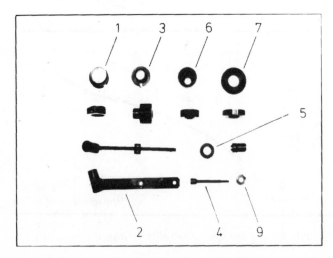

Figure 2.51
Workpiece
spectrum.

50

*Figure 2.52
Centring gripper for
cylindrical parts.
(Photograph: IPA).*

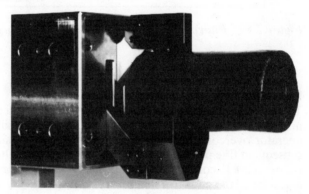

$$f(\delta)=\varepsilon=\frac{Z_1}{Z_2}=\frac{30}{23}=1{\cdot}3043, \delta=100°6'7'' \, (cf. \, Figure \, 2.34).$$

After examining possible gripping surfaces on the workpieces, gripping on the outer cylindrical shape proved to be the best possibility. The main problem was the diameter range from 6 to 60 mm, which was to be centred and gripped. According to value analysis, the gripper shown in Figure 2.52 was found to be the most suitable (see Figure 2.34 – three-point centring).

A systematic procedure which produces gripper designs for a wide range of workpieces is shown in Figure 2.53.

The development of a special flexible gripper which can be used for a workpiece spectrum is generally very expensive. Apart from this it is quite possible that no flexible gripper can be found by the method shown. Grippers are therefore being developed at present which are no longer restricted to a workpiece shape but adapt themselves automatically to a wide variety of shapes.

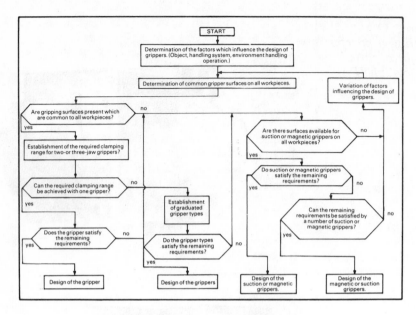

Figure 2.53   Plan of procedure for the rough design of grippers.

Resetting of grippers

The workpiece shapes found in practice vary considerably, as also do the intrinsic machine shape of workpieces within the permissible tolerances. The human operator overcomes these problems of shape by using his adaptable hands and subsequent fine positioning, which is controlled by the eye and the sense of touch, i.e. sensory. If common gripping surfaces similar in position and shape can be found for all the parts in a spectrum of workpieces, a gripper can often be developed by conventional means for the entire spectrum.

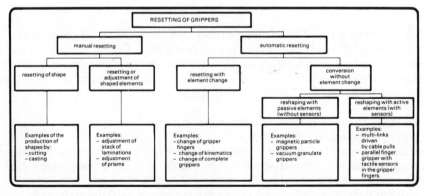

Figure 2.54   Resetting process for grippers.

If the number of similar gripping surfaces in the workpiece spectrum is insufficient, two problems arise:

1. Adaptation of the gripper surface to the workpiece shape to make a form-dependent clamping connection.
2. Moulding of the workpiece without exceeding permissible deformation values and causing changes in position.

These problems may be overcome by different methods (Figure 2.54). Manual resetting is at present the only method possible for many flexible handling systems for achieving shape adaptation. One or two automatic resetting processes have been proposed, but many of these are still being developed. Some, however, are admirably suited for industrial applications (e.g. gripper change devices).

Important automatic gripper resetting devices are represented in the following.

Gripper change-over

The cost of gripper change-over for automatic resetting depends on differences of shape and dimension within the workpiece spectrum. This cost relates not only to development and manufacture, but also to maintenance, storage and to the programming expenditure (Figure 2.55).

Figure 2.55
Expenditure for
change-over gripper
system as a function
of the number of
grippers.

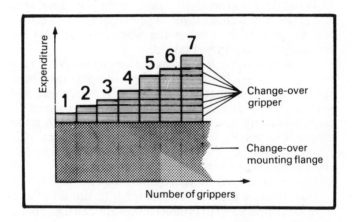

Figure 2.56   Gripper change-over system
for an industrial assembly
robot (Photograph: IPA).

53

Before a gripper change-over system is used it should therefore be determined whether it is possible at least to reduce the number of change grippers by increasing the possible applications for individual grippers. The increased planning expenditure resulting from this procedure will be compensated for by the lower maintenance and programming expenditures and, above all, by the lower proportion of the gripper change time in the total working cycle. On the whole it may be stated that gripper change-over systems do not involve any fundamental technical problems. Where there is a wide variety of shapes, these gripper change-over systems can, however, be very expensive. The design of the gripper change-over system developed at IPA Stuttgart (2.17) is shown in Figure 2.56.

Grippers with form-adaptable elements
In order to adapt gripper elements to workpiece shapes, active components may be used which are characterised by a control circuit with sensors and positioning drives.

Grippers with several fingers, analogous to the skeletal structure of the human hand, are representative of this type of gripper (2.18). The fingers are driven, for example, independently by cables. For control or regulation purposes, the motor current or the clamping force is measured, for example. However, gripper forms (2.19) which are structurally simpler have been developed, such grippers have only one drive, which closes three fingers together. The central gripper is guided by the tooling, so that even when the workpieces are not rotationally symmetrical a form-dependent grip is obtained.

It is also possible to use passive elements which have a form-adapting action without sensors. In this case, for example, the fingers are forced on to the workpiece, these fingers consisting of a diaphragm with a granular filling. The granular filling is displaced during the pressing process. In the gripper developed by Shinko Electric, the iron powder with which the gripper is filled is "fixed" by magnetic forces (Figure 2.57). In a process developed by IPA Stuttgart, the granular filling is "fixed" by vacuum (2.20). The advantage in the use of passive elements is the essentially simpler form adaptation compared with active elements. These grippers are therefore more economic than active form-adaptable grippers.

The object of these developments is to simplify the production of workpiece-adapted grippers and to produce uncomplicated gripper components which can be put to flexible use.

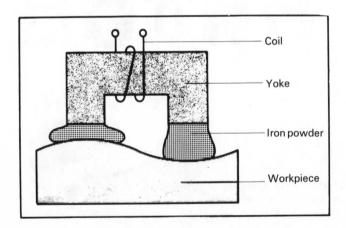

Figure 2.57
Diagrammatic illustration of the form-adaptable magnetic gripper from Shinko Electric (2.20).

Coil

Yoke

Iron powder

Workpiece

Development trends of flexible grippers

The trends in the development of flexible form-adaptable grippers which replace manual resetting are pointing in different directions. On the one hand grippers are being developed whose range of application is being widened by form-adapting elements or by increasing the clamping capacity and by additional functions (Figure 2.58).

The other possibilities being examined at the moment include gripper change-over systems. Only the future will determine which process will be preferred; probably both systems will be developed side by side and complement each other.

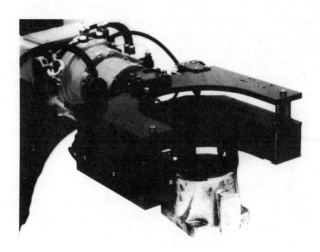

Figure 2.58
Three-finger
gripper with a very
large clamping
capacity (range).
(Photograph:
Siemens AG).

2.4.4.3   Design criteria for grippers
A   Calculation of the holding forces

In order to calculate the required holding forces, the following values must be determined or established:

1. workpiece weight
2. position of centre of gravity
3. size and position of gripping surface
4. maximum workpiece acceleration and its direction
5. insertion forces and their direction
6. safety coefficients
7. coefficient of friction between the workpiece and gripper surface

Workpiece weights and the position of the centre of gravity may be obtained or calculated together with the maximum workpiece acceleration (2g in most cases for controlled drives) and information on the position of the gripping surface. In addition to these forces, the insertion forces, which depend, among other things, on the accuracy and compliance of the devices and on the industrial robot, also must be taken into consideration.

According to whether the gripping is to be effected by form-dependent action or by pressure grip, the coefficients of the workpiece and gripper material must also be taken into consideration in order to determine the normal forces required to act on the gripping surfaces. The gripping forces calculated after considering the safety coefficients will act as a basis for designing the gripper components. However, it will be necessary in most cases to optimise the finished gripper, as a large number of estimated values must generally be included in the calculation. It is nevertheless important to carry out the calculation of the gripper forces as accurately as possible, since over-dimensioning because, say, safety coefficients are set too high, will lead to slower movements of the industrial robot or to larger amplitudes of vibration.

B Calculation of magnets

The calculation of magnetic grippers can be carried out by means of works data. In the calculation, two viewpoints are particularly important. On the one hand it should be assessed whether the gap between the workpiece surface and the magnet surface remains within sufficiently close tolerances, since magnetic forces are very greatly reduced if swarf, for example, widens the gap. On the other hand it must be remembered that the residual magnetism of magnets may lead to "sticking", and the use of electromagnets has therefore proved useful. They are not switched off to detach or loosen the tool, but a reducing a.c. voltage is applied, so that the residual magnetism is gradually reduced. Further possibilities are:

Detaching the tools from the magnet by plungers or by interrupting the magnetic flux by driven, displaceable yokes.

C Calculation of suction devices

Tables are also supplied by the manufacturers for calculating suction devices. It should be pointed out here that there are three different methods for applying a vacuum to suction devices:
1. vacuum pumps (Figure 2.59).
2. suction device with built-in Venturi nozzles which are operated by compressed air.
3. Suction cups which are released by compressed air (Figure 2.60).

These suction devices are pressed onto a workpiece and build up the vacuum because of their elasticity.

From the point of view of safety in the event of any energy failure, these suction cups are the safest, although their operational control is not as simple as in cups operated by means of a vacuum pump.

With suction grippers, leaks may also be deliberately accepted if the area of the opening is small enough to guarantee sufficient vacuum in connection with the vacuum pump. For small parts, e.g. gear-wheels, which are generally difficult to grip, suction grippers of this type are often used.

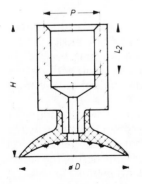

Calculated holding force (N) at 25% vacuum	Dimensions of the suction device (mm)				
	D	H	P	L₁	L₂
.7,8	20	25	R¹/₈"	10	7
43,0	50	33	R¹/₈"	10	7
173,0	100	50	R¹/₄"	10	9

Figure 2.59: Suction device and technical data. (Photograph: Carl Freudenberg.

56

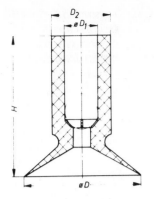

Holding force* F (N)	Dimensions of suction cups (mm)				Compressed air release pulse (bars)
	D	H	D_1	D_2	
3.0	6.3	12	2	4	1.0–1.4
4.5	10	13	3	5	1.0–1.4
10.0	16	20	5	8	1.0–1.4
30.0	25	30	8	12	1.0–1.4
60.0	40	45	12	20	1.0–1.4
140.0	63	60	20	32	1.0–1.4

*Gripping surface ground $R_{max} \leqslant 5\,\mu m$

Figure 2.60   Suction cup and technical data (Photograph: Carl Freudenberg).

D   Calculation of toggle levers

In designing toggle levers four lengths may be varied (Figure 2.61). *Lever length I_4* should be as short as possible to minimise the torques acting on the gripper. The other three levers generally have the same length, as this is the most space-saving design.

Figure 2.61
Toggle lever with its characteristic lengths I_1 to I_4

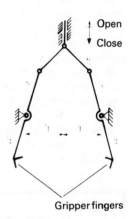

The gripping force for levers of ideal rigidity approaches infinity at the dead centre. The actual gripping force that can be achieved depends on the classic deflection of the kinematics (2.12). The levers should be as rigid as possible to ensure that the moment effect (high moment ratio-high gripping forces) may also be obtained. The high force amplification obtained with toggle levers can only be achieved within a very small clamping range, with the result that only workpieces with small tolerances can be securely gripped.

The calculation of toggle levers will in most cases be limited to a determination of the bending stresses of levers I_3 and I_4, because of the difficulty in calculating the overall elasticity, and in this case the forces used as a basis for the calculation must be assumed. Testing of the toggle levers is therefore absolutely essential.

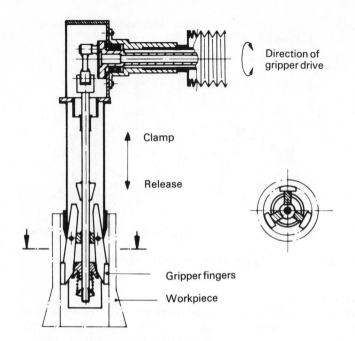

*Figure 2.62
Gripper for
centrifugally cast
pipes (according to
(2.13)).*

2.4.4.4 Special problems in application

Special problems in the use of grippers may arise because of the environmental stresses, particularly *thermal stresses*. An example of this is the handling of red hot forgings or hot pieces of glass. There are several possibilities of reducing the thermal stress on grippers. Firstly, the grippers can be immersed in a cooling medium during the handling cycle. Another possibility is to cool the gripper internally, e.g. by means of a cooling water circuit (often used for welding guns). For glass handling asbestos suction cups are used. These exhibit high thermal stability and sufficient insulation.

The use of high temperature materials and the removal of heat-sensitive components from the hot zone is generally very important (Figure 2.62). Another problem often arises in inserting of parts, if the pre-positioning of the parts to be inserted is not sufficiently accurate. In this case it is often possible, in the case of a

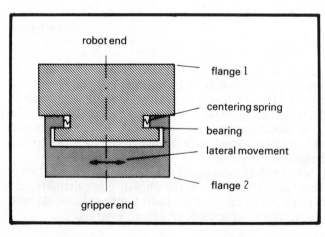

*Figure 2.63
Example of the
basic design of a
floating gripper
support.*

vertical insertion axis, to provide the gripper with a degree of *float* so that the insertion can be carried out safely. (Figure 2.63).

When loading machines using industrial robots it is sometimes advisable to use a double gripper to *shorten the idle times.* The double gripper consists of two grippers actuated independently, which can interchange their positions either by rotation or linearly (Figures 2.64 and 2.65).

One of the grippers handles the unmachined part, the other the finished part. In this case a workpiece can be extracted from the clamping device of the machine tool with one gripper and a new workpiece fed in with another gripper without any substantial change in the position of the handling system and with a consequent reduction in handling time.

Where several workpieces are machined simultaneously by a machine tool similar advantages are afforded by the use of *multiple grippers* (Figures 2.67 and 2.68).

Another problem may arise in *compensating for positional gripper errors* which

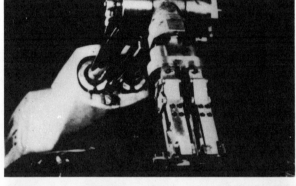

Figure 2.64
Double gripper on a
Unimation industrial
robot with swivel
movement for position
change.

Figure 2.65
Double gripper on a
loading device.
(Photograph: ZF AG).

Figure 2.66
Multiple gripper for
removal of body parts
from a press.
(Photograph: VW AG).

occur when loading chucks. These positional errors may be eliminated by the use of compliant elements in the gripper. If workpieces have to be turned, e.g. when facing both ends, *additional axes of rotation* are incorporated in the gripper (Figures 2.68 and 2.69).

Machining operations may also be carried out in the gripper, e.g. cutting off the sprues of die-castings (Figure 2.70).

Figure 2.67
Multiple gripper for
armatures. (Photograph:
Robert Bosch GmbH).

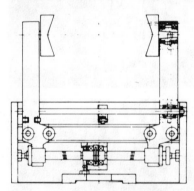

Figure 2.68   Special gripper with
integrated axis of
rotation for loading
machine tools (2.21).

Figure 2.69.   Gripper for turning over body parts. (Photograph: VW AG).

60

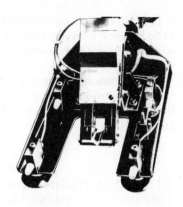

Figure 2.70   Suction gripper with integrated sprue cutter. (Photograph: Kaufeldt AB).

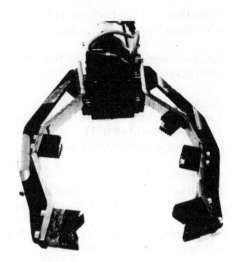

Figure 2.71   Gripper finger for square blanks 40 to 70 mm, and disc-shaped finished forgings up to 300 mm diameter. (Photograph: Kuka).

Figure 2.72   Gripper design with integral linear axis. (Photograph: Robert Bosch GmbH).

61

In *multimachine servicing,* the problem often arises that different workpieces are to be handled with one gripper. If it is not possible to position the gripping surfaces so that they are the same for all the workpieces, special solutions must be found (Figure 2.71).

To perform insertion movements, it may be necessary to integrate *linear feed movements* into the gripper. This will be necessary if the main axes movements of the robot are too coarse or if the working area, e.g. within the machine tool is too confined (Fig. 2.72).

2.4.5 List of suppliers of grippers
The gripper suppliers listed do not supply industrial robots.

Albert Fezer, 7300 Esslingen 11, Postfach 5024
– Vacuum pumps, suction devices –

Androidengesellschaft GmbH u. CoKG, Rossdorf, 06154/9093
– Parallel grippers, finger grippers –

Baer Automated Systems Inc.
Lakeland Industrial Park, Route 4, Box 1200,
Lakeland, Florida 83803
– pneumatic finger grippers –

Carl Freudenberg
Simritwerk, Abtlg. TNP, Postfach 1380, 6940 Weinheim
– pneumatic finger grippers, suction cups, suction devices –

Claus Polack, 7000 Stuttgart-Degerloch, Leonorenstr. 17
– Venturi suction devices.

Festo Pneumatic
7300 Esslingen 1 (Berkheim), Riuter Strasse, 0711/3911
– Venturi suction devices

Livernois Automation Company
25200 Trowbridge, Dearborn, Michigan 48123
Tel. 313-278-0200
– Vacuum grippers, pincer grippers (specially for press feeding) –

Sommer Automatic, Kaiser-Friedrich-Str. 148,
7530 Pforzheim, 07231/42194
– Venturi suction devices –

In addition some of the industrial robot manufacturers and suppliers may be able to provide grippers.

3. Testing of Industrial Robots

On examining the often very extensive technical descriptions of industrial robots, it will be seen that in most cases a wide variety of terms is used for the technical data. If we try to obtain more exact information it is soon recognised that different things are understood by the same terms, due on the one hand, to the fact that these terms are not yet standardised because of the newness of this equipment (terms from other fields, e.g. in the case of NC machines, cannot be transferred without difficulty), and on the other to the fact that different methods of measurement are used for specifying the technical data. There is therefore a need to develop measurement methods and acceptance guidelines based on them, in order to achieve standardisation of the characteristics of industrial robots and thus to make the different units comparable.

For this reason it is logical to test industrial robots according to a test programme. The measurements on industrial robots may provide information on interrelationships between equipment design (e.g. kinematics, type of drive) and equipment characteristics (e.g. positioning errors, rigidity). Manufacturers of industrial robots should be given the opportunity, on the basis of measured results, to improve existing equipment or to optimise future developments. Users should be enabled to assess industrial robots now on the market on the basis of these criteria, to compare them and to make the optimum use of them to solve their problems.

All the characteristics of industrial robots are given in tabular form in Chapter 3.1 on the basis of a test programme. The tables may serve as a basis for measurements of industrial robots and as a recommendation for equipment acceptance. The use of the test programme is explained in a short description of the most important measurements and by means of examples of measurement results.

Since position measurements in three coordinates still present problems, possibilities are indicated in Chapter 3.2 for constructing suitable measuring devices, even using simple fixtures and inexpensive sensors, to enable both manufacturers and users to test industrial robots at least on the basis of the most important geometrical values.

The possibility of a qualitative comparison of industrial robots on the basis of the requirements dictated by the specific applications is indicated in Chapter 3.3, together with prospects for further developments in the testing of industrial robots and the results of these tests.

3.1 Test programme for industrial robots

The test programme and associated methods of measurement are primarily user-

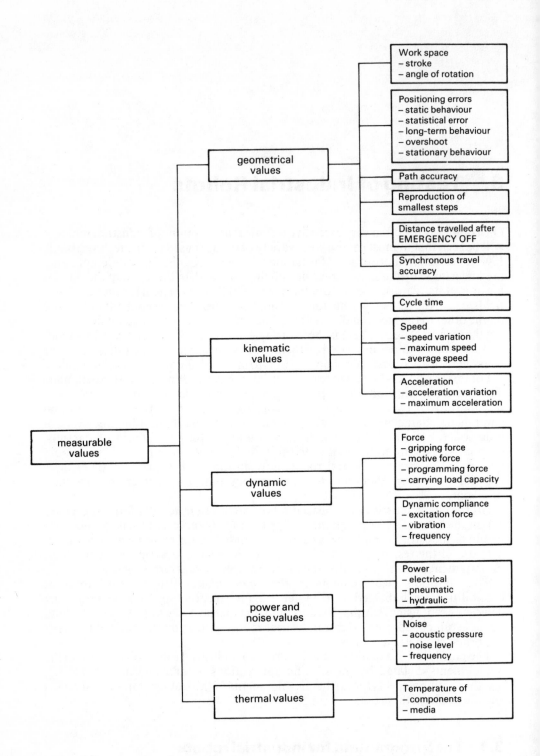

Figure 3.1   Test parameters for industrial robots.

oriented, i.e. the industrial robot should be tested by the means available to the user. Measures taken with regard to the control or measuring system, or equipment modifications, are not advisable. Special test operations, e.g. examination of vibrational behaviour for structural improvements in components, are mentioned, but they do not form part of the measurements described. Since the technical characteristics of the different industrial robots vary over a wide range, it is worth drawing up and developing a universally applicable test programme so that specials, e.g. simple feeding devices can be designed according to a problem-adapted subprogramme. The framework of the general test programme covers such a wide range that all the equipment at present on the market can be covered according to its geometrical, kinematic and dynamic values.

Figure 3.1 shows the test parameters for industrial robots. The parameters to be considered, such as:
– stroke,
– load and
– speed,
appear similarly in the test values, so that the test programme is represented in the form of a matrix. The grading of the parameters depends on the type of equipment and application, and is established for individual cases before a test, since the varying of all the parameters would increase the measurement cost to an extent which would no longer be justifiable. In describing the test programme, the parameters of a known device are used as examples. The percentages for the test parameters relate to the maximum value specified by the manufacturer.

3.1.1 Geometrical values
As an explanation of Figure 3.2, a definition is also given of what is understood by a "cycle" and "test cycle". In a "cycle", a position is approached several times without the motion as such being taken into account or established. Only one position should be reproduced. Positioning errors should be determined by this method. A "test cycle" is understood to mean several standardised cycles in which, for example, all the axes should be moved in a defined sequence, or, if feasible, simultaneously.

The "test cycles" should incorporate the most difficult test conditions for the industrial robot. In such a "test cycle", all the parameters should be covered if possible, such as maximum stroke, movements at maximum acceleration and speed, positioning at maximum and minimum extension.

The manufacturer's specifications for workspace are checked without load. In this case there must be a clear separation between main, auxiliary and tool axes (Figure 3.3). The non-utilisable movement area, which is in most cases missing from manufacturing specifications, is also determined, because it must also be taken into consideration by the user in allowing for collisions. The ratio of the "working area" to the "non-utilisable area" provides information on the suitability of the kinematic structure.

An examination of the workspace is necessary since the theoretically determined workspace frequently does not correspond to the actual space because of structural tolerances. Also of interest is the difference between the "mechanical" and "control" workspaces found in many devices. The possibilities of collision of the equipment with itself or other surfaces must also be examined. No special measuring accuracy is required. Figure 3.4 shows the workspace of an industrial robot which has been programmed by guiding the equipment (e.g. industrial robot for paint spraying).

What is very clear here is the deviation at the outermost limits of the working

Description			Measurement value: Name	Measurement value: Unit	Individual values statistically recorded	Static Cycle Test cycle	Parameters: Stroke %	Parameters: Load %	Parameters: Speed %	Test objective
Geometrical values		Workspace	Strokes and angles in all axes	mm	Individual values	static	0-100	0	–	Checking of manufacturer's specifications
Geometrical values	Positioning errors	Static behaviour	3-D path variation	mm	Individual values	static	0 50 100	0, 20, 50, 100, 150	–	Systematic error in loading elastic behaviour
Geometrical values	Positioning errors	Statistical error	3-D path variation at any point	mm	$\bar{x}$, s from 10 measure-ments	Cycle	0 50 100	0 50 100	10 100	Position variation hysteresis
Geometrical values	Positioning errors	Long-term behaviour	3-D drift of a point	mm	recorded	Cycle	50	100	100	Initial and long-term behaviour with accompanying temperature measurement
Geometrical values	Positioning errors	Overshoot	3-D amplitude outwards	mm	recorded	Cycle	50 100	0 50 100	50 100	Safe distance, periodic time
Geometrical values	Positioning errors	Stationary behaviour	3-D position variation at rest	mm	recorded	static	50	50 100	–	Control movements
Geometrical values		Path accuracy	2-D position variation along a path	mm	recorded	Test cycle	0-100	50 100	50 100	Path deviation in CP control systems
Geometrical values		Reproduction of smallest steps	Smallest path that can be travelled	mm	Individual values	static	–	50	–	Smallest step
Geometrical values		Distance travelled after EMERGENCY OFF	Distance travelled	mm	Individual values	Cycle	50 100	50 100	50 100	Examination of reaction time and displacement after EMERGENCY OFF
Geometrical values		Synchronous travel accuracy	3-D position variation during movement	mm	recorded	Test cycle	50	50 100	20 50 100	Checked synchronous travel accuracy of ("Tracking")

Figure 3.2 Geometrical values in the test programme for industrial robots.

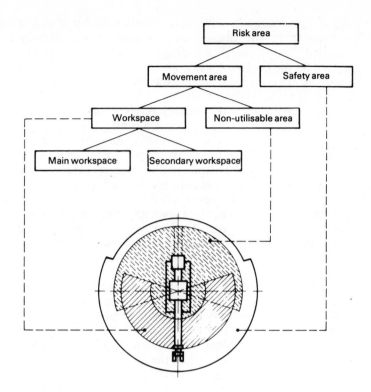

Figure 3.3
Classification of
the risk area
of industrial
robots.

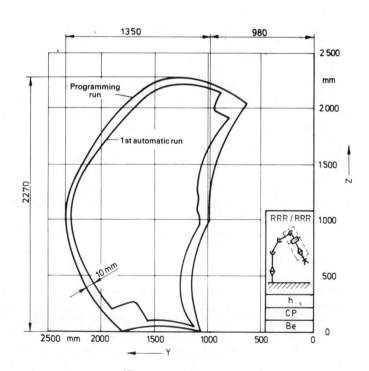

Figure 3.4
Vertical
section through
the workspace
of an industrial
robot.

area. These "jumps", which are clearly noticeable in the automatic cycle, are attributable to the measurement ranges in the measurement systems being exceeded. This means that points can be reached by the kinematics which are not covered by the control system. This fact may be of interest to users of the systems, since they might possibly wish to travel the workspace to its limit, but cannot reach this point in automatic operation.

Another reason for the measured deviations is the load relief of the tool in automatic operation, since the programming forces are eliminated. This cause of error, however, does not only arise at the limits of the workspace.

The **static behaviour**, for different loads and displacements, may also be termed positioning error. The position of an industrial robot set in an inside position may necessitate a correction, for example, because of the greater deflection in an outer position. In the measurements, a distinction may be made between measurements at the gripper for different extensions, and measurements on the frame at maximum extension. Using the first method, relationships may be recognised between extension and load, which might possibly be used for error compensation. In the second measurement method, weak points of axes, guides and joints become obvious.

The typical bahaviour of an industrial robot of **articulated arm design** is shown in *Figure 3.5*. In the loading and curve (M1), 3 kinks are seen due to bearing clearance. As the load is increased step by step the initial reaction is elastic. When the load exceeds a certain value the static friction in the bearing is overcome and the bearing clearance is taken up. Since the load is distributed unevenly, this takes place in the bearings at different times. When the load is relieved, only the elastic deformations are removed so that the initial position is not reached. In this case the last axis can be clearly located as the weak point by plotting the load curves M 2 and M 3.

The **static error** is the positioning error generally indicated by the manufacturers. On the basis of the error data for numerically controlled machine tools, a distinction should be made, as shown in Figure 3.6, between positioning tolerance T_{EU}, positional variance R_{PU}, hysteresis value U, position deviation A_U, standard deviation s and mean value $\bar{X}_i$.

The essential difference between industrial robots and machine tools is that the position set point is not stated numerically. The first value measured (programmed point) is used as the set point, and therefore incorporates an error already. The measurements to determine static errors are carried out ten times at several points in the workspace, with variations of load and speed. The measurements are carried out in the temperature stabilised condition, i.e. when temperature-dependent deviations are no longer observed.

Figure 3.7 shows the results of measuring the positioning error of a pneumatically driven industrial robot with fixed stops. The rotation unit was moved and the deviation from the theoretical position measured in directions X, Y and Z. The designation of the axes relates to the axis of a 3D measuring instrument used for this purpose (see Chapter 3.2).

The **long-term behaviour** should provide information on the time required to achieve thermal stability, i.e. until temperature-dependent positional deviations are no longer observed. This test, which is of particular interest as far as hydraulic units are concerned, lasts several hours in most cases. For the manufacturer of industrial robots, relationships are recognisable between temperatures of components and the resultant positioning error. The user is given information on warming up time and effects of interruptions such as breaks or breakdowns.

Figure 3.8 shows the temperature changes of components of an hydraulically driven industrial robot, measured over a period of four hours. The temperature

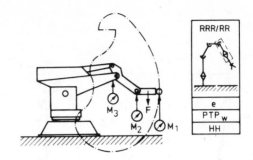

*Figure 3.5
Loading and unloading
curve of an
industrial robot
with articulated
arm design.*

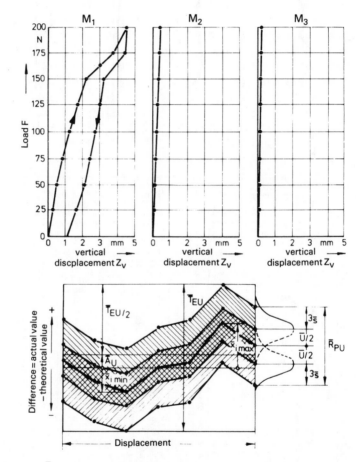

*Figure 3.6
Determination of
the static
characteristics
including hysteresis
(to VDI-3254) (3.1).*

$\overline{T}_{EU}$ = mean positioning tolerance

$\overline{R}_{PU}$ = mean positional variance

$\overline{U}$ = mean hysteresis range

$\overline{A}_U$ = mean positional deviation

$\overline{s}$ = mean standard deviation

$\overline{X}_i$ = arithmetical mean of the mean values $\overline{x}_i$ from both approach directions

69

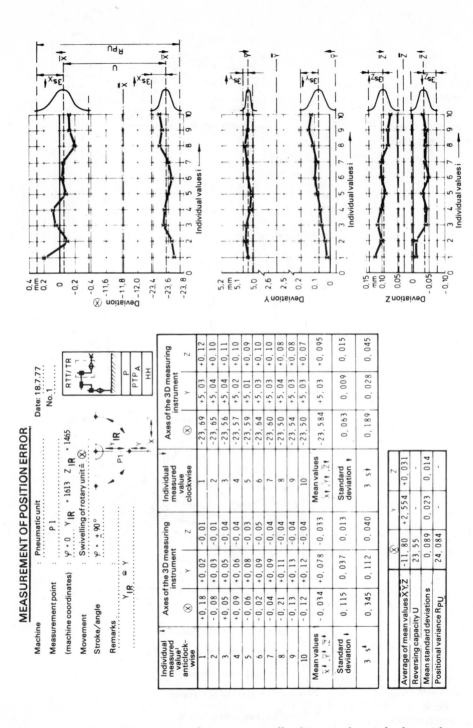

Figure 3.7   Positioning errors of a pneumatically-driven industrial robot with fixed stops.

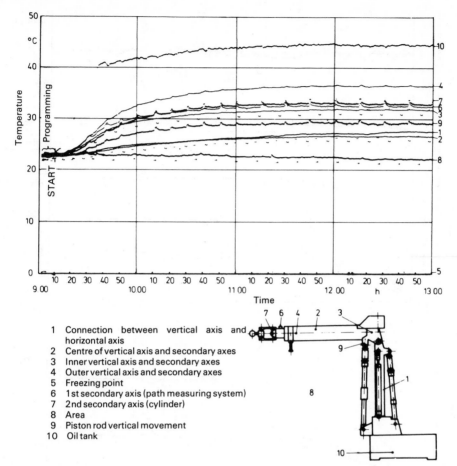

Figure 3.8   Temperature changes of the components of an industrial robot during
a long-term test.

1 Connection between vertical axis and horizontal axis
2 Centre of vertical axis and secondary axes
3 Inner vertical axis and secondary axes
4 Outer vertical axis and secondary axes
5 Freezing point
6 1st secondary axis (path measuring system)
7 2nd secondary axis (cylinder)
8 Area
9 Piston rod vertical movement
10 Oil tank

measurement is carried out in conjunction with the measuring of the positioning deviation in 3 coordinates, during which the movement of the industrial robot must be interrupted for a short time if the temperature is determined by contact measurement. The interruptions can be seen on the temperature curve.

In order to obtain similar conditions for the long-term mesurements for different units test cycles are established in which all the axes are moved. To arrive at a qualitative comparison, the following parameters must be equal:
– movement sequence,
– displacements,
– loads,
– speeds,
– measuring conditions,
– measuring cycles (interval between 2 measurement),
– time required for carrying out the measurements (for contact measurement) and
– number of test cycles between 2 measurements for different sizes.

These requirements can only be met for sizes and values with the same kinematics and the same overall size, and which have adjustable speeds. In the case

71

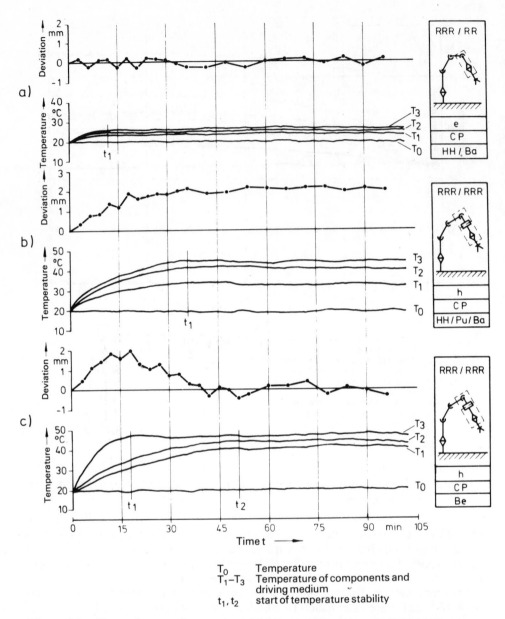

Figure 3.9   Typical curves for positional deviation of industrial robots in the long-term test.

of the equipment and measured results described below, the same measuring conditions were maintained. The results therefore only provide minimal quantitative information. However, the trend of the results allows conclusions to be drawn on the system examined. Measurements of the long-term behaviour of several industrial robots led to the conclusion that there is a direct relationship between temperature development and position deviation. Since the increase in temperature is attributable to the drives, and to a lesser degree to friction in the guide elements, a comparison of machines with different drives would seem appropriate. Figure 3.9 shows typical long-term curves for different units. Figure 3.9a shows the long-term drift of an electrical unit, measured over several hours. A maximum temperature rise of 5° C was measured on the equipment, but this did not result in any change of position.

The deviations indicated represent a statistical variation which was confirmed by comparisons with the statistical error, which was determined by 10 measurements within a short space of time (approx. 15 min.).

It is not possible, however, to make the general statement that there is no long-term drift in equipment with electric drives. However, in units which exhibited a much greater temperature rise than described above, a long-term drift could be demonstrated which lay within the positioning error specified by the manufacturer. In hydraulically driven equipment the clearest relationship is seen between temperature increase and positional deviation owing to the temperature transmission of the driving medium oil to components of the equipment.

In pneumatic equipment, which is in most cases positioned by adjustable stops, only slight temperature changes occur (cooling) due to expanding air, e.g. at cylinder inlets. No long-term drift is observed. In the case of pneumatic drives which are controlled by externally supplied energy (e.g. by electromechanical brakes), temperature-dependent positional deviations are observed because of the resultant heat generation.

Figures 3.9a and b show the typical behaviour of units which develop high temperatures. If all the components reach their maximum temperature at about the same time (T1 in Figure 3.9b), positional stability can be demonstrated from this time onwards within the statistical error. If components which affect the positional deviation reach their maximum temperature at different times (T1 and T2 in Figure 3.9c), an error trend may be observed in the opposite direction, which leads , as in the case under consideration, to an error compensation, but which may also lead to an error of opposite sign. Longer stoppages (approx. 20 mins), which are characterised by a marked reduction in temperature, also give rise to positional deviation. In principle the following recommendations may be derived from these measurements:

– avoid temperature increases (cooling),
– reach the operating temperature (T1) quickly (preheat) and keep constant (control),
– allow the equipment to warm up, and
– avoid protracted stoppages.

It should also be noted at this point the the long-term errors measured for some units lay within the tolerances specified by the manufacturer and that for many applications the measured deviations can be allowed for.

Overshoot
In industrial robots oscillations may occur if violent changes in direction take place, or if the unit is accelerated or retarded. Generally speaking, free oscillations occur, i.e. after a position is reached, no energy is supplied to the system. Since resistances such as bearing friction, air resistance and inner friction occur during the movement, the oscillation ceases automatically after a certain time, i.e. the oscillations are damped.

In order to examine the operating behaviour, an analysis of the overshoot is necessary, with interest centred on the maximum oscillation amplitude and the

Oscillation wave (basic)	Direction	Contributory Factors	Causes
a) 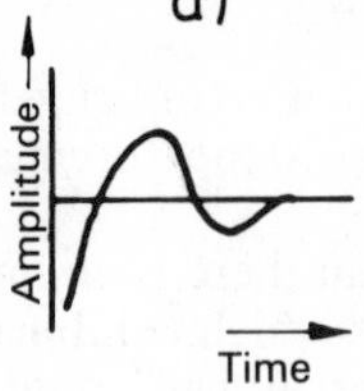 *[graph: Amplitude vs Time]*	Behaviour similar in all directions	Amplitude and frequency dependent on ● speed ● load ● overhang	● built-in springs and cylinders act as damping elements ● generally during rotation due to high moment of inertia
b) 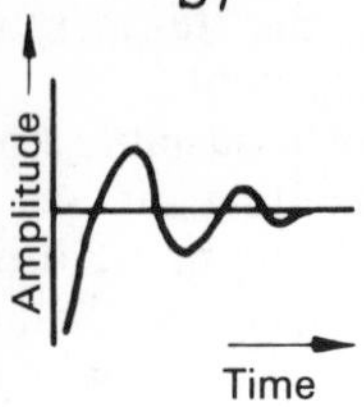 *[graph: Amplitude vs Time]*	Behaviour similar in all directions	Amplitude and frequency dependent on ● speed ● load ● overhang	● no outside exciters ● no feedback ● attenuation of the oscillation due to internal forces (friction) ● especially typical for equipment with fixed stops
c) 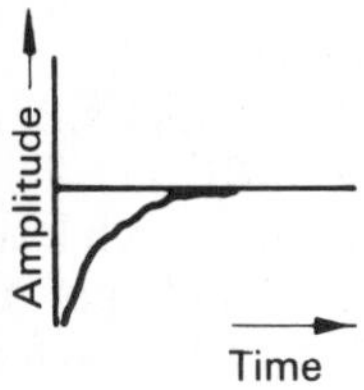 *[graph: Amplitude vs Time]*	only in the dimension in which the main movement takes place	● control characteristics ● speed amplification	● speed control ● position control ● especially typical of equipment with computer control
d) 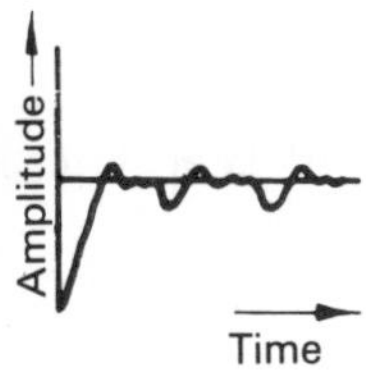 *[graph: Amplitude vs Time]*	only in the direction of the force of gravity	● load ● overhang	● transient pocess is a damped osciallation ● reduction due to force of gravity ● subsequent adjustment to theoretical position

Figure 3.10   Typical oscillation waves of industrial robots.

time taken for a value below a certain permissible level to be reached.

In order to determine the overshoot of an industrial robot on reaching a position, feeding is effected into a non-contact 3-D measuring head with a measuring cube. The attenuation is plotted 3-dimensionally, and the test repeated at different points in the working area, at different loads and speeds.

Typical oscillation characteristics for industrial robots are shown in Figure 3.10. The oscillation values (e.g. damping) not only provide information on drives, control system and measuring system, but also on positioning error (amplitude)

74

and periodic time, which is included in the cycle time as a user-relevant value (Figure 3.11).

The following recommendations may be derived from the measurements of the **damping behaviour**:

For the user, the maximum amplitude and **decay time** are of interest as important cycle time values. The cycle time can often be reduced by controlled approach to the end position or by approaching at a reduced speed, thus avoiding oscillations. When approaching intermediate positions, e.g. when bypassing obstacles, no allowance need be made for oscillations.

The oscillation wave with all the measurement parameters (factors) should be shown by the manufacturer, or at least the time taken for a value below a certain permissible amplitude to be reached.

A determination of the **path accuracy** is only appropriate for industrial robots which are provided with a control system suitable for following a path (CP). However, straight lines may also be examined by the proposed method of measurement in the case of PTP machines. For metrological reasons, the path is only measured in 2 dimensions, and the **path curve** and the reproduction of the paths are shown (Figure 3.12).

During the measurement, a steel scale is arranged in the workspace so that the starting and end points of the path curves show different values along all co-ordinates. By approaching a measuring head with the equipment a defined distance between sensor and scale is obtained (see Chapter 3.2.1). This position of the measuring head is programmed as the initial position of the path.

In order to obtain the same distance from the scale at the end point of the path, the equipment is moved in steps until the signal from the sensors corresponds with the initial signal. To determine the path accuracy, the equipment is run in the automatic mode. Because of load variations between the individual cycles, deviations appear similar to those which appear in the measurements of positioning accuracy. These deviations also show up with static loads. As the speed varies, so errors occur due to **trailing** distances and **overswing**.

Figure 3.13 shows two curves with the same load applied to the gripper for each, but with different speeds. In plotting curve 1, v_1 was = 0.04 m/s, curve 2 v_2 was = 0.15 m/s. On comparing the curves, two different error influences can be seen. On the one hand a variation of the deviations about a theoretical path, on the other the

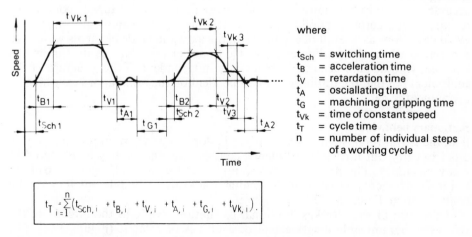

where

t_{Sch} = switching time
t_B = acceleration time
t_V = retardation time
t_A = osciallating time
t_G = machining or gripping time
t_{Vk} = time of constant speed
t_T = cycle time
n = number of individual steps of a working cycle

$$t_T = \sum_{i=1}^{n}(t_{Sch, i} + t_{B, i} + t_{V, i} + t_{A, i} + t_{G, i} + t_{Vk, i}).$$

Figure 3.11   Composition of the cycle time of an industrial robot.

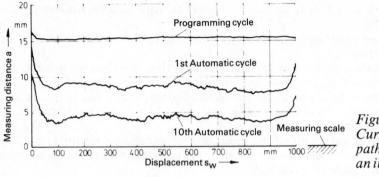

Figure 3.12
Curves of the
path accuracy of
an industrial robot.

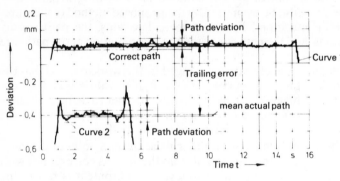

Curve 1: Path speed $v_1 = 0.04$ m/s
Curve 2: Path speed $v_2 = 0.15$ m/s

Figure 3.13
Path accuracy
of an industrial
robot.

deviation of the average actual path from the correct path. The variation in deviation produces the path error, which is brought about mainly by the **excess oscillation** of the gripper.

The deviation of the average actual path from the correct path is due to the **trailing** error, and is therefore speed-dependent. Just as in measuring positioning accuracy, the equipment must be brought to operating temperature by continuous running over a certain period before commencing the measurements, in order to avoid deviations as a result of expansion of the axes and guide elements which are superimposed on the trailing error as additional errors. Various measurements have shown that path accuracy is generally independent of the structure of the equipment but that the trailing error is dependent on the number of axes which go to produce the path.

Stationary behaviour

If an industrial robot with **actuated** drives is stopped in a certain position, e.g. in order to carry out further adjustments, control movements may be observed in many machines. The measurements should monitor these changes in position. Control movements can be avoided by suitable adjustment or different design of the control system (Figure 3.14).

In the case of shot strokes, frictional oscillations come to the fore. Here the **shortest programmable displacement** of a unit is of interest (Figure 3.14). For infinitesimal movements it must be expected that hysteresis phenomena will

76

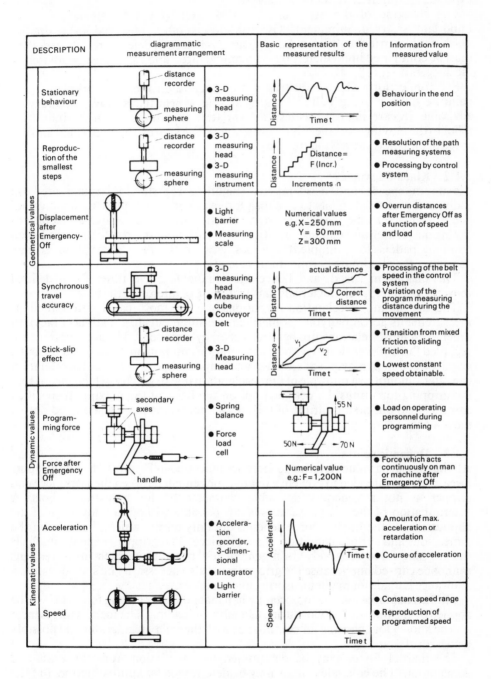

DESCRIPTION		diagrammatic measurement arrangement		Basic representation of the measured results	Information from measured value
Geometrical values	Stationary behaviour	distance recorder, measuring sphere	● 3-D measuring head	Distance / Time t	● Behaviour in the end position
	Reproduction of the smallest steps	distance recorder, measuring sphere	● 3-D measuring head ● 3-D measuring instrument	Distance / Distance = F (Incr.) / Increments n	● Resolution of the path measuring systems ● Processing by control system
	Displacement after Emergency-Off		● Light barrier ● Measuring scale	Numerical values e.g. X = 250 mm Y = 50 mm Z = 300 mm	● Overrun distances after Emergency Off as a function of speed and load
	Synchronous travel accuracy		● 3-D measuring head ● Measuring cube ● Conveyor belt	actual distance / Distance / Correct distance / Time t	● Processing of the belt speed in the control system ● Variation of the program measuring distance during the movement
	Stick-slip effect	distance recorder, measuring sphere	● 3-D Measuring head	Distance / v_1 / v_2 / Time t	● Transition from mixed friction to sliding friction ● Lowest constant speed obtainable.
Dynamic values	Programming force	secondary axes	● Spring balance ● Force load cell	55 N / 50N ← → 70 N	● Load on operating personnel during programming
	Force after Emergency Off	handle		Numerical value e.g.: F = 1,200N	● Force which acts continuously on man or machine after Emergency Off
Kinematic values	Acceleration		● Acceleration recorder, 3-dimensional ● Integrator ● Light barrier	Acceleration / Time t	● Amount of max. acceleration or retardation ● Course of acceleration
	Speed			Speed / Time t	● Constant speed range ● Reproduction of programmed speed

Figure 3.14 Condensed representation of the measured results of industrial robot characteristics

77

become so marked that no reproducible initial movement actually takes place with the existing input signal. Thus for short strokes the positioning uncertainties to be expected are different for those for long strokes.

The checking of the **over-run after EMERGENCY OFF** operation is an important safety requirement. Overrunning is often observed in hydraulic machines in particular, often at full pressure. When such defects are detected, they must be eliminated immediately by the manufacturer, and users must be advised of the dangers (Figure 3.14).

For applications where industrial robots operate on moving workpieces (e.g. painting on a moving belt), checking of the **synchronous travel accuracy** is advisable. In carrying out this test, the variation in distance from the industrial robot to the moving workpiece is recorded, while the load on the equipment and the speed of the belt is varied (Figure 3.14).

3.1.2 Kinematic values
Maximum **speeds** and **accelerations** (Figure 3.15) should be measured on individual industrial robots to establish test cycles in which these extreme values occur. Damage to the workpiece due to inertia forces or excessive accelerations during the picking and placing operations is of particular interest to the user. When handling pallets, for example, the forces between the parts and the pallet may determine the maximum permissible acceleration.

The maximum speed, and hence the attainable cycle times for a defined movement cycle, are characteristic data which are of interest to the user.

The speed and acceleration variations are in most cases determined as measurements accompanying geometrical measurements (e.g. path accuracy), in order to establish relationships between accuracy and speed. The accelerations are recorded three-dimensionally, as the acceleration components at right angles to the actual direction of movement are often also important; e.g. tangential acceleration can often amount to a multiple of the acceleration due to gravity, which means that the gripping force must be designed not on the basis of the maximum specified carrying load but on a multiple of this, if safe holding is to be guaranteed during movement.

3.1.3 Dynamic values
All the measured values relating to force are indicated in Figure 3.16. The **gripping force**, which is intended to prevent a relative movement between the workpiece and gripper, is not a generally definable characteristic for industrial robots. A determination of the gripping forces is not possible without a knowledge of the application, as the holding force depends not only on the weight, but also on the dimensions and surface condition of a workpiece. The gripper shape itself affects the gripping force decisively. In a general test programme such measurements cannot be carried out because the gripper must also necessarily be specified unless allowance is made for major measuring uncertainties.

The **inertial force** may be important when machining or **mating** forces have to be applied by the industrial robot during assembly or loading. The inertial forces must also be known for checking for compliance with the safety regulations and possibly for design of guarding.

The inertial forces may be determined by calculation from the mass and acceleration. The equivalent mass may be determined by approximation, but the acceleration is measurable. In robots with pneumatic drive, a static mesurement is possible, i.e. the equipment can be held with a small force.

Programming forces may be measured in machines which are programmed by

78

Description			Measured value		Individual values	Static	Parameters			Test Object
			Name	Unit	Statistical Recorded	Cycle Test cycle	Stroke %	Load %	Speed %	
Kinematic values		Travel time	Time of stroke or cycle	s	statistical	Test cycle	–	0 50 100	20 50 100	Working cycle, cycle time
Kinematic values	Speed	Speed variation	3-D acceleration integrated	m/s	recorded	Cycle	50	0 50 100	20 50 100	Analysis of working cycles. Reproduction of speed, cycle time
Kinematic values	Speed	Maximum speed	3-D acceleration integrated	m/s	recorded	Cycle	100	0 50 100	100	Checking of manufacturer's specifications
Kinematic values	Speed	Average speed	Distance and time measurement	m/s	statistical	Cycle	50	0 50 100	20 50 100	Checking of manufacturer's specifications, cycle time
Kinematic values	Acceleration	Acceleration variation	3-D acceleration	m/s^2	recorded	Cycle	50	0 50 100	20 50 100	Representation of control behaviour, decay
Kinematic values	Acceleration	Maximum acceleration	3-D acceleration	m/s^2	recorded	Cycle	100	0 50 100	100	Acceleration peaks, e.g. for determining the gripping forces

Figure 3.15 Kinematic values in the test programme for industrial robots.

Dynamic values		Description	Measured value		Individual values Statistical Recorded	Static Cycle Test cycle	Parameters			Test Object
			Name	Unit			Stroke %	Load %	Speed %	
Force		Gripping force	Force between gripping surfaces	N	Individual values	static	-	-	-	Determination of the holding forces, application-specific
		Inertial force	Force, aceleration or impulse	N m/s²	Individual values or recorded	static or test cycle	100	100	100	Forces on collision, (operational safety) Machining forces
		Programming force	Force	N	Individual values	static	0-100	-	-	Determination of the forces to be applied for programming in "teach in"
		Carrying load	Force	N	Individual values	test cycle	100	0-150	100	Checking manufacturer's specifications
		Excitation force	Force	N	Recorded	static	-	-	-	Determination of dynamic rigidity
Dynamic elasticity		Oscillation	Amplitude	mm	Recorded	static	-	-	-	Determination of - dynamic elasticity - damping
		Frequency	Number of oscillations	Hz	Recorded	static	-	-	-	- resonance frequency

Figure 3.16   Dynamic values in the test programme for industrial robots.

guidance. If large forces must be applied by the programmer to move the industrial robot, this has two consequences. First, the quality of the path and hence the quality of the work suffers, e.g. in spray painting, where tedious trial and error and frequent repetition will be necessary. Secondly, loads are exerted on the kinematics by the programming forces; these loads are not present in automatic operation and can therefore lead to additional path errors.

The **carrying load**, as this is itself a measured parameter, is determined from the results of the geometrical and kinematic measurements; i.e. a test must be made to determine the load at which the specified values can still be reliably obtained with regard to positioning error and average and maximum speed, without overloading the machine by excessive oscillation or increased temperatures. In special applications, not only the weight, but also the dimensions of a load are important, as with large loads an industrial robot is subjected to considerable torsional stress. However, incorporation in a general test programme is not advisable.

By contrast to the static behaviour of industrial robots, which is determined by unladen weight and load, and is influenced by the moments of inertia of the components and by bearing and joint clearances, the dynamic **rigidity** is determined by inertia forces, damping and weight distribution. For most industrial robots, the dynamic behaviour is only of interest as the grippers are able to oscillate at high amplitudes in the extended position if the dynamic rigidity is insufficient. Greater importance is attached to these tests when industrial robots are used for machining operations with contact with the workpiece, as in deburring. In this case the industrial robot is dynamically excited by the tool, like a machine tool.

Measurements and analyses of the dynamic behaviour are sufficiently well known from machine tools. For handling equipment, the elasticity at the gripper at different excitation frequencies is of particular interest.

The gripper of the unit is in this case excited dynamically by an oscillator. This force effects a change of path. The force and path signals are indicated on an oscilloscope and evaluated in a Fourier analyser. The elasticity frequency response curve may be represented as an amplitude and phase sequence or as a polar curve (Figure 3.17). From this can be determined dynamic elasticities, damping and resonance frequency (3.2, 3.3).

3.1.4 Power and noise values
Power and noise measurements (Figure 3.18) are only made as auxiliary measurements. They serve to determine peak values and mean values in continuous

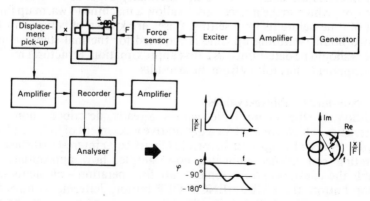

Figure 3.17 Dynamic elasticity measurement for industrial robots.

81

operation. In the case of hydraulically driven industrial robots, for example, designs are conceivable which are connected to the existing pressure oil circuit of a production unit. In these cases, pressure and volumetric flow must be measured.

For pneumatic machines, it is often only the inlet pressure in bars which is indicated, but not the volumetric flow. In a test programme the power must be determined so that the users are able to see how many of these machines can be connected to an installed network, or what power must be made available. In addition, the compressed air consumption is an important cost factor.

The power taken from the electrical mains is recorded with a wattmeter, and from this can be determined the average electric consumption. The mechanical efficiency may be determined by approximation from accelerations and moved masses. The power loss may be calculated roughly from the amount of dissipated heat; in the case of hydraulic units this is done by means of the temperature difference between the cooling water inlet and outlet and the volumetric flow.

There are detailed guidelines for noise measurements on machines (3.4, 3.5), in which measurement points are stated with reference to the outer surface of the machines. For handling equipment, a distinction must be made between units in which the kinematics are separate from the drive unt, and units where both are integral. In the former case the drive unit may be regarded as an independent machine.

The measurements may be carried out with reference to the surface. In the second case it is advisable to carry out the measurements outside the workspace, as access is permitted during operation. In this case the measurements are therefore carried out at the usual distances from the outer boundary of the workspace. Owing to the complexity of this method of measurement and the lack of objective information provided by acoustic measured values, the noise level measurements are restricted to those for which standard values are given in official regulations (3.6).

3.1.5 Thermal values

With these measurements, temperatures of components (drive, control system) of the handling equipment, the environment and, if appropriate, the hydraulic oil, are recorded (Figure 3.19). Temperature changes cause thermal deformations of the components and therefore positioning errors, or they affect the starting behaviour. Temperature variations are also recorded in the continuous tests as auxiliary measurements, so that the relationships between temperature and other measured values can be analysed (see also Section 3.1.1).

What interests the users here are the thermally dependent deformations after switching on, which make it necessary to allow a machine to warm up for a certain amount of time before its final setting. Cases are known of equipment for which about two hours' warming up time is required for thermal stabilisation. These influences should be determined as a first approximation. Structural modifications to the components may follow from these results.

3.1.6. Non-measurable test values

In addition to the quantifiable values already described, non-measurable characteristics, which are used for a qualitative assessment of industrial robots and frequently represent a high cost factor, are of particular interest to the user.

Operating convenience indicates how simple, how appropriate and how accessible the controls and instruments are for operation – elements such as the start-stop button, the EMERGENCY OFF button, lettering, control lights, the presence of a screen with plain text indication.

Power and noise values		Description	Measured value		Individual values statistically recorded	Static Cycle Test cycle	Parameters			Test object
			Name	Unit			Stroke %	Load %	Speed %	
Power		Electrical	Voltage, current	VA	Mean value Max. value Recorded	Test cycle	0-100	50 100	20 50 100	Power balance Operating costs, Connection data
		Pneumatic	Pressure Flow	Nm³	Mean value Max. value Recorded	Test cycle	0-100	50 100	20 50 100	Operating costs Connection data
		Hydraulic	Pressure Flow	VA	Mean value Max. value Recorded	Test cycle	0-100	50 100	20 50 100	Power balance, Connection data
Noise		Acoustic pressure	Pressure	$\mu N/m^2$	Recorded Max. value	Test cycle	100	100	100	Determination of noise emission;
		Noise level	Level = logarithmic ratio of values	dB	Recorded Max. value	Test cycle	100	100	100	Safety regulations
		Noise frequency	Frequency	Hz	Recorded	Test cycle	100	100	100	

Figure 3.18   Power and noise values in the test programme for industrial robots.

Description	Measured value		Individual values statistically recorded	Static Cycle Test cycle	Parameters			Test object
	Name	Unit			Stroke %	Load %	Speed %	
Temperature	Temperature variation	°C	Recorded	Test cycle	50	100	50 100	Temperature variation of media and components accompanying long-term measurements.

Figure 3.19   Thermal values in the test programme for industrial robots.

83

In connection with operating convenience **programming convenience** is an important value which may influence the usable time of an industrial robot. In this case the instruction time denotes the time which an operator requires to master the machine, at least for planned applications, including programming, setting, re-setting, elimination of simple errors, error diagnoses and maintenance. The time is conditionally comparable if a test cycle is used as a basis for different industrial robots, and if the same operator carries out this work.

Programming expenditure (time to prepare the first programme and time required for the first setting up) and re-setting (time for programme change and any gripper change) may also be compared on the basis of different standard cycles in which the given application must be used as the starting point.

Ease of maintenance indicates how frequently maintenance work must be carried out and how quickly it can be completed (e.g. dependence on accessibility, number of items to be maintained). In cost analyses of the use of industrial robots (3.7, 3.8), it has been shown that the maintenance costs must be set very high. Schiefelbusch (3.9) calculated 18% of the total costs for the repair and maintenance costs in one application, as against 5% in conventional production. A reduction of these costs, which are mainly cuased by breakdowns, can be achieved by:

● systematic detection and evaluation of the breakdowns on the premises of the user,
● rapid fault diagnosis in the industrial robot,
● rapid fault removal.

The possibility of fault diagnosis and removal may be established as part of the test programme, regardless of the application (e.g. fault indication, replacement of components).

In order to reduce the downtimes of industrial robots due to breakdowns, preventive maintenance must be aimed for, and this should be carried out outside operating hours. A further possibility lies in replacing the whole unit, but this is not possible in most cases because of the tolerances of components, or is fraught with difficulties. An examination of the **interchangeability** of similar equipment types forms part of the test programme.

In applications of industrial robots under particularly extreme conditions the units should be tested for their **resistance**, and in doing so only the effect of temperatures can be simulated before use (e.g. by heating the control system).

The **life** of industrial robots can only be related to application experiences over a lengthy period.

In the case of new equipment, it is necessary to determine how far **safety regulations** are complied with (generally VDE, VDI guidelines). In addition, collision possibilities, EMERGENCY OFF conditions going beyond general regulations, and possibilities of incorrect operation must also be examined.

3.2 Methods of measurement for testing industrial robots

3.2.1 Measuring heads for geometrical measurements

The most important test values are the geometrical values. For all geometrical values the problem is the measuring of a position or the deviation from a position 3-dimensionally as accurately as possible. An analysis of possible displacement measurement methods (ultrasound, capacitive, inductive, etc.) has shown that purchasable inductive displacement pick-up closest to meeting the requirements

Description	Contacting		Non-contact			
	Touch-trigger probe	3-D measuring head	3-D measuring head with 3 sensors	3-D measuring heads with 6 sensors	2-D measuring head with 2 sensors	2-D measuring head with 4 sensors
Diagrammatic representation						
Direction of contact	Distributed arbitrarily over hemisphere	X, Y, Z	X, Y, Z	X, Y, Z	Y, Z	Y, Z
Contact force	1.5 N (at contact) 2.5 N (over-travel)	1÷3 N	–	–	–	–
Stroke or measuring distance	7 mm over-travel	20 mm	10 mm	10 mm	10 mm	10 mm
Weight	1 kg	1 kg	1 kg	1.2 kg	0.5 kg	1 kg
Referenzkörper — Description	Cube with notch, 50 mm	Sphere, 50 mm, any	Cube, 100 mm		Straight edge, 100 mm, 1500 mm	
diagrammatic representation						
Material	any	any	metal		metal	
Mounting	Probe on 3D measuring instrument. Reference body on robot	Head on 3-D measuring instrument, reference body on robot	Head on 3-D measuring instrument. Reference body on robot.		Head on robot. Ruler on measuring plate.	
Contact			Distance a	Distance $a \sim \dfrac{a_1 + a_2}{2}$; Angle $\alpha \sim \dfrac{a_1 - a_2}{b}$	Distance a	Distance $a \sim \dfrac{a_1 + a_2}{2}$; Angle $\alpha \sim \dfrac{a_1 - a_2}{b}$

Figure 3.20 Measuring heads for measuring geometrical values of industrial robots.

85

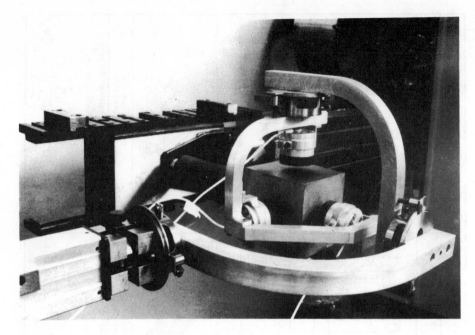

Figure 3.21   Non-contact 3-D measuring head with reference cube.

with regard to working distance, resolution and linearity. Using these displacement pick-ups, 3-dimensional measuring heads (3-D measuring heads) may be constructed for position measurements, and 2-dimensional measuring heads for path measurements, which are sufficiently accurate. In principle a 3-D measuring unit with a touch-trigger probe, which, on contact, supplies an electrical signal (Figure 3.20), is also suitable.

Basically a distinction can be made between contacting and non-contact measuring heads. The advantage of non-contact measuring heads lies in the fact that the reference unit, which is fixed to the robot gripper, can be fed into the head

Figure 3.22
Reproduction of the path of an
industrial robot.

Figure 3.23
Test stand with industrial robot and
3-D measuring instrument.

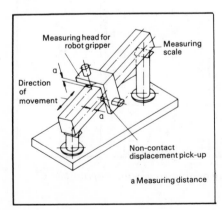

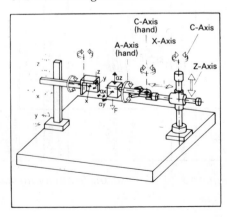

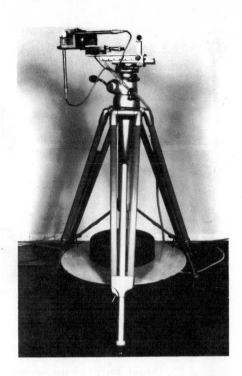

Figure 3.24
Tripod for positioning 3-D
measuring heads.

without measurements having to be interrupted. A further advantage is that the 3-dimensional decaying of oscillations can be measured simultaneously in one position. For path measurements where a relative movement between the measuring head and the reference body is a precondition, only non-contract measurement methods can be used.

Figure 3.21 shows a non-contact 3-dimensional measuring head and the reference cube, which is secured in the gripper of an industrial robot. The 3D measuring head can be rotated about 3 axes and is mounted on gimbals, which permit any orientation of the displacement in the area without the point of intersection of the pick-ups varying relative to the 3-D measuring instrument.

In the case of continuous path controlled industrial robots, measurements must be carried out to determine how accurately a programmed path is reproduced. A different measurement method is provided for measuring the reproduction of a programmed path (Figure 3.22).

A measuring head with 2 non-contact displacement pick-ups is secured in the gripper of the industrial robot. In the programming operation, this measuring head is guided along the measuring scale at a defined distance, and the curves are plotted. The distance curves of the automatic cycles are then plotted and compared with the programming curve. The following of a particular path is not necessary as a straight line which runs diagonally in space is also a "path" for an industrial robot, since it must travel along all the axes simultaneously. In this way the measurement of a "path" can be considerably simplified.

3.2.2   Test stands for geometrical measurements
In order to survey industrial robots throughout the workspace within the required range of accuracy, accurate alignment is required between the industrial robot and the measuring equipment, which can be achieved by means of a test stand whose

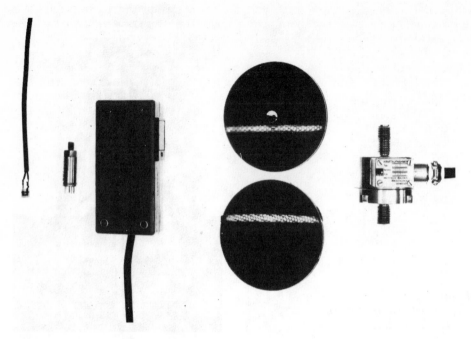

Figure 3.25   Equipment for non-geometrical measurements.

mechanical part consists of a measuring plate and a 3-dimensional measuring instrument (Figure 3.23).

Such a test stand provides a means of positioning measuring heads quickly and accurately in the workspace. For the use of contact measuring methods, this is essential. Furthermore, there is the possibility of standardising the measured results with regard to the parameter "stroke", since the position of the measurement point is known from an indication of the 3-D measuring instrument relative to a reference point. A simpler possibility of positioning a measuring head is provided by a tripod to which a slide is fitted carrying the measuring heads (Figure 3.24). One advantage of this system is its mobility, which also permits measurements at the point of application of the industrial robot, but one disadvantage is the unknown position of the measuring heads.

3.2.3   Measuring equipment for non-geometrical values
In addition to the 3D measuring heads and the 3D measuring instrument, equipment may be used for non-geometrical measurements, the most important of which are briefly described (Figure 3.25).

Thermocouples
The thermocouples should have an adhesive base and be designed for all surface temperature measurements on industrial robots. Because of the minimal thickness of the thermocouples, there is a guarantee that the actual surface temperature will be recorded, and this can be improved even further by the application of thermal paste.

The thermocouple elements are made from copper and constantan foil. Their voltage characteristic conforms to DIN series 43710. The conductors are inserted between two cresylic resin foils, and the thickness of the thermocouples is approximately 0.1 mm. The thermocouples are flexible, so that they can easily be adapted to a particular surface.

Light barrier
A light barrier is recommended for calibrating speed and measuring the EMERGENCY OFF movement. It uses a reflector, with transmitter and receiver incorporated in one housing. Thus electrical wiring need only be fitted on one side.

Load cell
The force is measured by displacement. The measuring principle in this case is the same as for an inductive contact displacement detector with a stronger spring. The inductances of two coils, connected as a half-bridge, vary as the displacement varies. By choosing different springs different force ranges can be measured.

Accelerometer
Accelerometers are used for measuring both the acceleration of moving objects and oscillating and shock accelerations.

The **accelerometer** consists of a mass suspended in a housing by a spring and its movement relative to the housing is damped by a liquid of constant viscosity.

Externally acting accelerating forces displace the spring-supported mass in the acceleromter and this acts as an armature between two inductances which are connected as an **electric half-bridge**.

The accelerometer is calibrated by tilting through 180°. The difference in readings between these two positions shown is $2\,g = 19.62$ m/s^2

3.3 Summary

3.3.1 Comparison of the measured results
The conclusions derived from the measurements given by way of examples in preceding sections, where they relate to statements regarding the quality of a machine, cannot easily be applied to similar equipment as each unit exhibits its own characteristics due to the interaction of all the factors which determine its operating behaviour. Nor do the pre-conditions exist for a quantitative comparison of the equipment characteristics, as the measuring conditions can never be exactly the same and most characteristics cannot be given as a single numerical value.

Nevertheless the possibility of assessing limited comparability will be indicated in the following on the basis of a comparison of the measured results from different industrial robots (Figure 3.26).

An assessment of unit characteristics according to the values "good", "average", "poor", is based on the application for which the equipment is intended. For this purpose all the test results which must be provided for each unit in the form of measurement series, curves and diagrams must be used.

A decision with regard to suitability must be based on the requirements which depend on an examination of the workplace.

Universally applicable statements on common features of similar types of equipment can, as shown in Figure 3.26, only be made with regard to the drive on the basis of long-term behaviour, and with regard to the kinematics on the basis of static behaviour, with individual units again making exceptions.

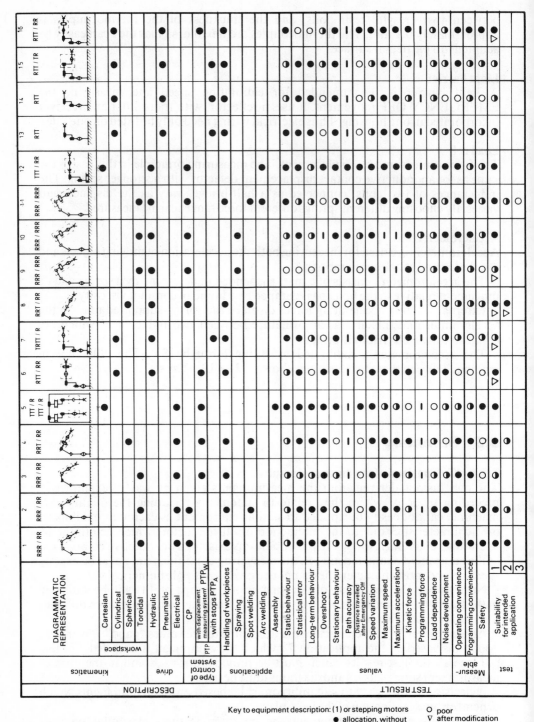

Figure 3.26   Summary of assessment of test results from industrial robots.

A presentation of test results by this method must not be used under any circumstances to assess a unit without analysing the detailed individual results.

In order to be able to assess the suitability of an industrial robot for a particular application additional measured results obtained under working conditions should be used, together with an evaluation of breakdown statistics spanning several years. However, these results so far only exist for a few units.

3.3.2 Conclusions from testing industrial robots

The test programme specially developed for industrial robots includes all the measurable characteristics of a machine. By the use of recommended measuring procedures and instruments it is possible to measure industrial robots so that unambiguous, reproducible values are obtained for each unit, which enable it to be described, compared and used correctly in a predetermined application according to its characteristics.

The basic conclusions drawn from the measurements of different industrial robots are summarised in the following (3.10):

● A quantitative comparison of the measured results from different industrial robots can only be made with difficulty. The precondition for comparability of the measured results is that the same measurement methods and the same measurement parameters, such as speed (maxima and variations), load, position, and the same measuring cycles (distance and time) must be used. All the conditions can be fulfilled only where the units are identical.

In the final analysis, the requirement regarding identical speed variations cannot be met.

● A user-specific comparison of different units is possible:
The evaluation of the individual measurements and an assessment based on requirements arising from workplace analyses permit a statement to be made on the suitability of a machine for a projected task. To obtain a final assessment, these measurable characteristics (mainly geometrical values) must be supplemented by an evaluation of the non-quantifiable characteristics. By establishing standard tasks, these characteristics are also measurable to a limited degree, such as programming convenience on the basis of time required to program this task. Valuable additional information is provided by data gained from experience (e.g. statistics on breakdown data on industrial robots that have been operating for some considerable time.

● Universally applicable statements on equipment types on the basis of one of their distinguishing characteristics cannot readily be made:
Robots are generally classified according to one of their characteristics (e.g. drive, kinematics, measurement system, control system).
As the behaviour of a unit during operation is determined not by one of these factors alone but by the combination of all the factors, each unit has its peculiar characteristics which qualify it to a greater or lesser extent for a specific task.

● Numerical values alone do not provide sufficient information, they must be complemented by their changes with respect to time:
The operating behaviour of a unit cannot be represented by comparing measured values from different units, but rather by determining the variation of the measured values (e.g. in the extended life test) and by comparing different mesured values (e.g. displacement with temperature). The same applies to an analysis of the weak points of a unit.

In future, the testing of industrial robots will increase in importance in connection with equipment acceptance and in determining weak points of the equipment. Where measured results are made quickly available, the possibility is provided of optimum adjustment of the equipment for a specific application. Simplified test methods will provide the possibility of carrying out measurements at the workplace according to reduced test programme. This will ensure the best possible use of the equipment and early detection of shortcomings.

4. Feeding devices

4.1 Terms used in handling engineering

Whilst terms and designations from production engineering may be regarded as widely known, special terms in handling engineering are at present only known to a relatively narrow circle of specialists. The terms "feeding functions" and "work-piece group" will therefore be explained in the following.

4.1.1 Feeding functions

The functions performed by feeding devices may be broken down into individual basic functions – the feeding functions. A feeding device generally performs several feeding functions simultaneously or successively, but always carries out at least one feeding function.

The feeding functions and their associated symbols are suitable, among other things, to give a clear representation of complicated handling processes. Functions performed in duplicate can be detected quickly and avoided with the use of these symbols.

The feeding functions are represented in Figure 4.1, in accordance with VDI guideline 3239 – symbols for feeding devices:

On close examination of these feeding functions, however, it will be seen that on the one hand they partly overlap while on the other hand without additional information they only have a limited information content. Thus the functions feed-in, feed-out and transfer must be regarded as the equivalent of the "execution" of a translatory movement. In this case there is no reference to any boundary conditions to be satisfied, e.g. with regard to distances to be covered, accuracies and speeds required for the construction of a feeding device. The VDI feeding functions are therefore only suitable for the qualitative description of a handling process.

The functional breakdown of the handling process (4.2), with the feeding functions required for it, which are briefly explained in the following, is represented in Figure 4.2.

Storage:
Storage does not represent an active, but a passive process. It must not be confused with the loading and unloading of parts into or out of a storage unit.

Changing orientation and position:
Physically, "changing orientation" may be regarded as "turning about one to three axes", and "changing position" may be regarded as "moving in space", and there-

93

Feeding function	Symbol	Description
Binning		Binning is the random storage of workpieces in containers suitable for them, to build up stocks before and after operations.
Magazining		Magazining is the storage of workpieces in a definite orientation.
Transfer		Transfer denotes both the orientated and the unorientated flow of working material. The movement may be caused either by gravity or by forced movement.
Branching		Branching is a form of transfer. It denotes a segregation of workpieces from an orientated or unorientated workpiece flow, in order to divide this flow up or sort out a workpiece.
Combining		Combining is a form of transfer and denotes the combining of workpieces to form a workpiece flow.
Ordering		Ordering brings workpieces from any position into a predetermined orientation and position.
Orientation check		Orientation check establishes a workpiece orientation for later placing in location.
Turning over, rotating, swivelling		Turning over, rotating, swivelling, bring workpieces into a new orientation and/or direction.
Metering		Metering denotes the segregation of a certain number of pieces or quantity of material at a point of operation.
Feed-in		Feed-in is the direct and controlled movement of the working material into the point of operation.
Positioning		Positioning is the creation of an exact workpiece position.
Clamping		Clamping is the fixing of the workpiece in a certain position.
Releasing		Releasing denotes the release of the clamping force.
Feed-out		Feed-out is the removal of the workpiece from the point of operation.
Machining		Operation carried out at operating points.

Figure 4.1   Summary of feeding functions.

Figure 4.2 Functional
breakdown of the handling
operation.

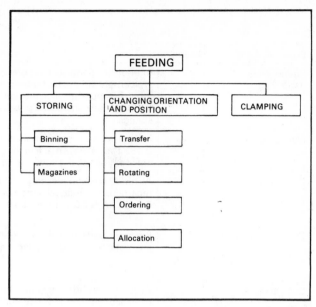

fore correspond to the execution of rotational and translational movements respectively. Orientation and position may change, as shown in subdivided form in Figure 4.2, depending on the relationship between the initial and final positions of the objects being handled.

Clamping:
In conformity with the definition of the clamping system as a sub-system belonging to a processing system, the function "clamping" is defined as follows:

"Clamping means securing the position of an object in such a manner that it can easily be released".

4.1.2 Workpiece groups
In order to solve problems of workpiece handling systematically, a workpiece classification system has been developed, which divides the entire range of work-pieces into only 12 different groups. All workpieces which belong to the same group exhibit similar or identical properties, and hence also similar behaviour with respect to their handling under the influence of gravity. The individual basic groups are described in Figure 4.3.

These 12 groups, with additional information on the workpiece and handling characteristics (Figure 4.4), produce a 14-digit characteristic number via a code. This is to allow the use of systematic procedures for solving handling problems and the use of solution catalogues or an EDP system.

A detailed description of this classification scheme is given in the book "Handhabungseinrichtungen" (Handling Devices) (4.3).

4.1.3 Differentiation between feeding devices and transfer devices:
It is often difficult to make a clear distinction between a transfer device and a feeding device, as the same piece of equipment may be both a means of transfer and a feeding device, depending on the application.

This applies particularly to transfer devices. The following definition of feeding devices will help to clarify the differences.

Workpiece group	Example	Description
Interlocking parts		Interlocking parts are workpieces of different shapes, whose typical behaviour is that they can interlock. Assignment to this group is simple because the workpieces interlock immediately in the bunkered condition, e.g. lock washers, piston rings, helical springs.
Flat parts		Flat parts extend primarily in two dimensions ($0 \leq C/A \leq 0.49$). They always occur in the preferred flat position. They are of any shape but interlocking must not be possible.
Rotational parts		All rotational workpieces, without deviation of shape, are regarded as cylindrical parts, with a length/diameter ratio of $0.5 \leq C/A \leq 30$, e.g. plain bolts, shafts, rods or bars.
Prismatic parts		Prismatic parts are solid block-shaped workpieces with prismatic shape, e.g. angular, triangular, square.
Conical parts		All full cones and truncated cones, regardless of whether circular or not, oblique or straight, must be assigned to the "Conical part" group.
Pyramidal parts		Workpieces of regular and general pyramid shape are classified in this group, e.g. wedge, double wedge.
Headed parts		Headed parts are understood to mean all simply stepped workpieces of cylindrical, prismatic, conical or pyramid-shaped geometry, e.g. screws, rivets.
Hollow parts		Thin- and thick-walled hollow parts not enclosed on all sides, of cylindrical, prismatic, conical or composite shape must be allocated to the "hollow part" group, e.g. box sections, spacer sleeves.
Complex rotational parts		By complex rotational parts are meant workpieces which are stepped more than once, or which have form variations on non-curved axes, such as crankshafts, camshafts.
Irregular solid parts		Workpieces with curved and/or intersecting axes and of mainly solid shape, e.g. pressings and forgings.
Spherical parts		All spherical and approximately spheroidal workpieces are included in this group, e.g. ball bearing balls, etc.
Long parts		Solid, flat and hollow sections with $C/A \geqslant 30$, e.g. steel strip from the coil.

Table 4.3   Summary of workpiece groups.

96

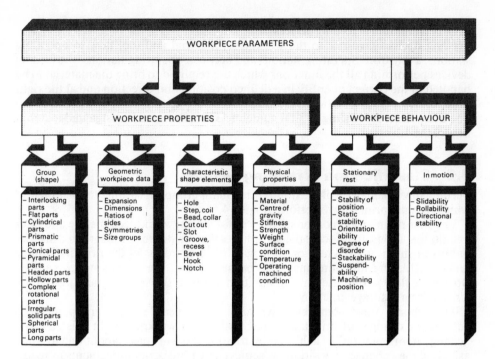

Figure 4.4   Summary of workpiece parameters.

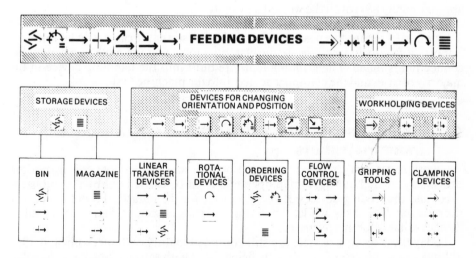

Figure 4.5   Classification of feeding devices.

4.1.4 Definition of feeding devices

Feeding devices give rise to the workpiece, material and tool flow to operating stations, away from operating stations and between operating stations, the feeding devices performing all the functions which are required to bring the material to be handled in the correct quantity in a defined position and direction and at the right time to the operating station, and to remove, store or transfer it at the end of the operating time. According to VDI guideline 3300, this area is designated material flow of the 4th order.

4.2 Classification of feeding devices

Feeding devices may be divided into three main groups according to the function which they are primarily to perform (Figure 4.5).

The first main group comprises devices for storing workpieces before and after operating stations. In this case the workpieces may be stored in a disordered fashion in bins or in an ordered fashion in magazines. Regardless of the primary storage function, feeding functions may also be installed in these units which cause a workpiece flow within the unit. In addition, units in the second main group may also perform a storage function.

The second, most comprehensive and most versatile main group includes devices for changes of orientation and position of workpieces. In this case the "change in orientation" is achieved by rotating the workpiece about one to three axes, and the "change in position" is achieved by translatory movements in space. These units are used to effect the workpiece flow into, out of and between operating stations.

It is clear from a summary of the most important and most frequently used feeding devices (Figure 4.6), showing which feed functions can be performed with a particular unit, that feeding devices and industrial robots have the most wide-ranging applications.

One of the most important handling functions, "ordering", however, cannot be performed by these handling units.

The ordering devices form a further major item in this main group, since the problem of ordering arises, according to a representative investigation (4.2), in approximately 88% of all the workplaces analysed.

The third main group includes gripping tools and clamping devices whose function is to secure the position of a workpiece so that it can easily be released. Important factors in this case are the clamping force and the method whereby the connection is made. The gripping tools are built into the handling device in the majority of cases and the clamping devices into the machine tool.

4.3 Requirements for feeding devices when used with industrial robots

At workplaces which are automated with industrial robots all operations previously carried out by human operators must be carried out by the industrial robot and the feeding devices. It is appropriate for the industrial robot to perform the pure handling tasks, whilst control and ordering tasks are taken over by feeding devices suitable for this purpose.

Today these feeding devices are still equipped mainly with mechanical baffles, although in some cases they are now being fitted with sensors. Mechanical baffles are used, among others, in vibratory bowl feeders. The parts are scanned only, and

Function / Device		≷	⇄	→	↦	↗	↘	→|	→	+|+	|→	⌒	≡
BINS	Tote bin	●											
	Stillage	●											
	Box pallet	●											
	Feed hopper	●		●	●								
MAGAZINES	Tube magazine			●									●
	Piston feed magazine												●
	Channel magazine			●									●
	Ratchet feed magazine			●									●
	Helical magazine			●									●
	Pallet magazine												●
	Chain magazine			●									●
	Drum magazine			●									●
TRANSFER DEVICES	Industrial robot			●	●	●	●	●	●	●	●	●	
	Pick-and-place devices			●	●	●	●	●	●	●	●		
	Belt conveyor			●									●
	Roller conveyor			●									
	Vibratory conveyor			●									
	Walking beam conveyor			●									
	Roller or slide conveyor			●									●
	Rail conveyor			●									
	Rotary indexing Table			●						●			●
	Reel			●	●								●
METERING DEVICES	Pusher			●	●	●	●	●			●		
	Gate				●	●							
	Diverter				●	●	●						
	Detent				●								
	Screw			●	●								
ORDERING DEVICES	Vibratory bowl feeder	●	●	●									
	Elevating hopper feeder	●	●	●									
	Magnetic feeder	●	●	●									
	Disc hopper	●	●	●									
	Rotary feeder	●	●	●									
	Rotating drum feeder	●	●	●									
	Centre board hopper feeder	●	●	●									
	rotary hopper	●	●	●									
	Separating devices for parts that can tangle	●	●	●									

Figure 4.6 Summary of the most important feeding devices.

possibly diverted, or they are scanned and transferred from an incorrect position to a correct position, according to whether the baffle is passive or active. One disadvantage of this, however, especially when set against the flexibility of an industrial robot, is that the mechanical baffles must always be adapted to the particular workpiece.

Feeding devices have therefore already been developed that are fitted with sensors which require no conversion when changing workpieces. The sensor signals are processed into commands (e.g. call-up of a certain sub-program) to the industrial robot.

A number of sensors suitable for recording the properties of workpieces and converting them to a suitable signal are summarised in Figure 4.7.

4.4. State of the art

The automation of workplaces has been limited, as already mentioned, mainly to the automation of the actual machining process. Only in places where the same workpiece is being continuously manufactured in large quantities has there also been partial automation of the handling operations. The machines required for this were always designed and built for a particular application. One of these rigid methods of interlinking is known by the term transfer line, where the machining stations are connected by special handling units designed only for this purpose and for this application.

However, a transfer line is not suitable for the automation of workplaces at which frequently changing workpieces are machined. For such work-places 'flexible' handling units are required, which can be easily adapted to different workpieces.

With the construction of industrial robots it has been possible to produce flexible handling units which are freely programmable and which perform practically every desired handling operation. One disadvantage of such units is that today, and

Type of sensor	Reaction to	Type of output signal
Optical sensors		
Light barriers	Interruption of light beam	Digital
Light curtain	Interruption of light beam	Digital
Light curtain, measuring	Interruption of light beam	Analogue
Photodiodes	Interruption of light beam	Digital
TV systems	Objects in the field of view	Digital
Optical correlators	Comparison with master pattern	Digital
Mech./Electr. sensors		
Probe	Mech. actuation	Digital
Switch	Mech. actuation	Digital
Inductive proximity sensor	Approach of metal	Digital, analogue
Capacitative proximity sensor	Approach of material	Digital, analogue
Magnetic switch	Approach of magnetic material	Digital
Piezo-electric devices		
Pressure transducers		
Force transducer	Compressive force	Analogue
Strain gauge	Strain	Analogue
Potentiometer	Rotary displacement	Analogue
Fluidic sensors		
Air orifice	Approach of object	Analogue
Jet-plate systems	Distance between jet and plate	Analogue

Figure 4.7   Summary of sensors.

in the foreseeable future, they are not nor will be provided with sensory facilities. This means, for example, that the goods to be handled must always be supplied singly and in an orderly fashion to a certain position known to the robot. The sorting, segregation and transfer operations required for this are performed by suitable feeding devices, but it will be quickly seen here that the feeding devices now available on the market are far less flexible when compared to an industrial robot. At present there are no units designed specially for ordering which can easily be reset for several different workpiece ordering types.

In order to solve these problems attempts are now being made to combine feeding devices with elements for workpiece recognition. With such feeding devices it is possible to arrange for the industrial robot linked to them to carry out a required locating process itself by gripping the workpiece and then bringing it to the desired position and orientation by rotation, turning over and/or swivelling. The feeding device merely conveys the workpiece in such a combination. There are sufficiently flexible feeding devices for the feed function "transfer", which can be obtained as standard units.

It is to be expected that conventional ordering devices, such as the vibratory bowl feeder, will still be used as an economic solution in rigidly linked systems, whilst well-known types of feeding devices, equipped with newly developed workpiece

	Feeding devices / Basic solutions	Entangling parts	Flat parts	Cylindrical parts	Prismatic parts	Conical parts	Pyramidal parts	Headed parts	Hollow parts	Complex rotational parts	Irregular solid parts	Spherical parts	Long parts	<10 (01)	10÷24 (02)	25÷62 (03)	63÷99 (04)	100÷249 (05)	250÷629 (06)	630÷1000 (07)	>1000 (08)	Rollable (in guide)	Slidable	Suspendable	Stackable	Stable	Not shock sensitive	Not fracture sensitive	Not surface sensitive	Ferromagnetic
Magazines	Chain magazine	○	●	●	●	●	●	●	●	●	◐	●	○	◐	●	●	●	●	●	◐	○				●	○				
	Drum magazine	○	●	●	●	●	●	●	●	●	◐	●	○	◐	●	●	●	●	◐	◐	○				●	○				
	Spiral magazine	○	●	●	●	◐	●	◐	●	◐	○	○	○	○	●	●	●	◐	◐	○	○	▲	▲				▲	▲	▲	
	Pallet	○	●	●	●	●	●	●	●	●	◐	●	○	●	●	●	●	●	●	●	◐									
	Tube magazine	○	●	●	●	◐	◐	◐	◐	◐	○	●	○	◐	●	●	●	◐	◐	○						▲	▲	▲	▲	
Transfer devices	Industrial robot	○	●	●	●	●	●	●	●	●	◐	●	○	◐	●	●	●	●	●	◐										
	Pick-&-place devices	○	●	●	●	●	●	●	●	●	◐	●	○	●	●	●	●	●	◐											
	Belt conveyor	◐	●	●	●	●	●	●	●	●	◐	◐	○	●	●	●	●	●	●	◐						▲				
	Roller conveyor	○	●	●	●	●	●	●	●	●	◐	◐	○	●	●	●	●	●	●	◐		▲					▲	▲	▲	
	Vibratory conveyor	◐	●	●	●	●	●	●	●	●	◐	◐	○	●	●	●	●	◐	○	○							▲	▲	▲	
	Walk'g beam conveyor	○	●	●	●	◐	●	●	●	●	◐	◐	○	●	●	●	●	●	●	◐										
	Rotary indexing table	○	●	●	●	●	●	●	●	●	◐	●	○	●	●	●	●	●	○	○										
	Roller or slide conveyor	◐	●	●	●	◐	◐	●	●	◐	◐	●	○	●	●	●	●	●	◐	○		▲	▲				▲	▲	▲	
	Rail conveyor	○	○	○	○	○	○	●	●	◐	◐	◐	○	●	●	●	●	●	◐	○		▲	▲				▲	▲	▲	
Ordering devices	Vibratory bowl feeder	◐	●	●	●	●	●	●	●	●	○	○	○	●	●	●	●	◐	○	○	○						▲	▲	▲	
	Elevating hopper feeder	○	●	●	●	◐	●	●	●	●	○	○	○	○	●	●	●	◐	○	○		▲	▲				▲	▲	▲	
	Magnetic feeder	○	●	●	●	●	●	●	●	●	○	○	○	○	●	●	●	◐	○	○							▲	▲	▲	▲
	Rot. disc hopper feeder	○	◐	●	●	●	●	●	●	◐	○	○	○	●	●	●	●	◐	○	○		▲					▲	▲	▲	
	Rotating drum feeder	○	◐	●	●	●	●	◐	●	◐	○	○	○	●	●	●	◐	○	○			▲					▲	▲	▲	
	Cent. brd. hopper feeder	○	●	●	●	●	◐	●	●	◐	○	○	○	●	●	●	◐	○	○	○		▲	▲				▲	▲	▲	
	rotary hopper	○	○	○	○	○	○	●	○	◐	○	○	○	●	●	●	◐	○	○	○					▲		▲	▲	▲	
	Separate device for parts which get entangled	◐	○	○	○	○	○	○	○	○	○	○	○	●	●	●	◐	○	○								▲	▲	▲	
Metering devices	Pusher	○	●	◐	●	○	○	○	○	○	○	●	○	●	●	●	●	◐	●	●					▲		▲			
	Gate	○	◐	◐	◐	◐	○	◐	◐	○	○	●	○	●	●	●	●	◐	○		▲	▲		▲		▲	▲	▲	
	Screw	○	○	○	○	○	○	○	○	●	○	●	●	●	●	●	◐	◐	○		▲					▲	▲	▲	
	Rollers, strips $V_2 > V_4$	○	●	◐	●	◐	○	○	◐	○	○	◐	○	●	●	●	●	◐	○	○				▲		▲	▲	▲	
	Transfer disc, drum	○	◐	●	●	◐	○	◐	◐	○	●	○	●	●	●	●	◐	○	○	▲	▲			▲		▲	▲	▲	

Table 4.8   Applications of the most commonly used feeding devices

recognition devices, are being used in a growing number of workplaces which are being automated with industrial robots.

4.4.1 Description of frequently used basic solutions

A large number of functional principles for feeding devices are known from the literature (4.4, 4.5), but only a few of the principles are used frequently. The reason for this lies first in the fact that these devices can only be used for a very limited workpiece spectrum, and secondly the solutions are in some cases uneconomic and technically complicated.

The applications of the most frequently used basic solutions are summarised and evaluated in Table 4.8. The evaluation is effected by using different symbols for individual criteria to indicate the evaluation: very good, neutral or negative.

The individual symbols denote the following:

○ not suitable

◑ conditionally or partially suitable

● suitable

The symbol (▲) shows the most important physical and behavioural properties which a workpiece must have for it to be suitable for a certain feeding device. A market analysis (Figure 4.9) shows that only a small number of feeding devices from this still relatively comprehensive range of basic solutions are obtainable as standard equipment on the market. The most frequently used basic solutions for the feed functions – binning – magazining – transferring – ordering – distributing, are explained in the following by means of a brief summary.

The remaining feed functions mentioned in VDI guideline 3239 may be regarded as sub-functions or as special cases of the above-mentioned functions.

The basic solutions were selected according to VDI guideline 3240. In order to provide a comparison between the individual basic solutions, the different function

	Feeding devices	Feed functions					Number Type	Overall sizes
Bins	Feed hopper	●		●	●		5	12
	Storage belt	●		●			1	1
Magazines	Chain magazine			●		●	7	14
	Drum magazine			●		●	3	7
	Spiral magazine			●		●	2	5
	Pallet magazine					●	1	2
Ordering	Vibratory bowl feeder	●	●	●			19	196
	Elevating hopper feeder	●	●	●			15	58
	Centre board hopper feeder	●	●	●			1	4
	Rotary disc hopper feeder	●	●				1	1
	Rotating drum feeder	●	●				3	7
	Separating device for parts which get entangled	●	●	●	●		2	5
Transfer	Belt conveyor			●		●	20	
	Roller conveyor			●		●	10	
	Vibratory conveyor			●			4	
	Walking beam conveyor			●			2	

Figure 4.9
Summary of feeding devices on the market

	Bin or hopper	
Description of function:	Bins and hoppers are containers in which workpieces are stored in an unordered state.	**Basic solutions**
Drive:	A distinction is made between fixed and moving hoppers, and hoppers with feeding elements. Moving hoppers and hoppers with feeding elements are driven electrically in most cases.	**Examples of design**
Control system:	Moving hoppers and hoppers with feeding discharge elements are preferably connected to linked installations with a control system.	
Resetting capability	Fixed and moving hoppers are suitable for all workpiece groups. In the case of hoppers with feeding elements, these elements must be adapted to the workpiece in question.	Discharge hopper with vibratory bowl feeder. (Photo: Blösch)
Design characteristics:	Capacity of the hopper. Space requirement. Type of feeding elements.	
Application:	Hoppers are used to store workpieces. Hoppers with a feeding device also transfer workpieces and/or sort or segregate the workpieces with suitable design of the feeding elements.	
Design forms:	(a) Fixed hoppers are closed containers with loading and unloading openings. (b) Hoppers with gravity feed. (c) Hoppers with unloading elements, e.g. vibratory conveyors, elevating conveyors, vertical conveyors, rotating drum feeders.	Distributing hopper (Photo: Feldpausch)

	Chain magazine	≡
Description of function:	Chain magazines are used for indexed transfer and storage of ordered workpieces. The workpieces are generally arranged in lines, horizontally or vertically. The workpiece holders are mounted on a rotating chain. The workpieces are positioned by nesting or clamping.	Basic solutions
Drive:	The drive is generally electric through a geared or step motor.	Examples of design:
Control system:	Chain magazines are generally connected in interlinked systems to an electrical control system.	
Resetting capability	Chain magazines are mostly designed for a particular workpiece. In specials they may also be adapted to different workpiece sizes for a particular workpiece group, within certain limits.	
Design characteristics:	Storage capacity. Resetting capability. Pitch. Space requirement. Fitting and installation position. Loading and unloading facilities.	Self-compensating magazine (Photo: Sormel SA)
Application:	Chain magazines are used in rigidly or loosely interlinked production lines. They can be closely adapted to the space available.	
Design forms:	Curtain magazine. Elevator magazine Container magazine.	Workpiece transfer table (Photo: Heyligenstaedt)

104

	Drum magazine	☰
Description of function:	The workpieces are arranged one behind the other on a magazine drum, i.e. on a circular surface in one or more rows. The axis of rotation may be horizontal or vertical. The workpieces are positioned in nests or are clamped. For loading and unloading the magazine is indexed.	**Basic solution:**
Drive:	The drive may be electric, pneumatic or hydraulic.	**Examples of design**
Control system:	Drum magazines are equipped with a control system.	
Resetting capability	Where workpiece cassettes or interchangeable workpiece holders are used, the drum magazine may be reset for similar workpieces.	
Design characteristics:	Storage capacity. Resetting capability. Space requirement. Fitting and installation position. Loading and unloading facilities.	Feeding device with circular magazine (Photo: Diedesheim)
Application:	Drum magazines may be used on individual magazines and in interlinked systems.	
Design forms:	Circular magazine. Circular indexing table. Ring magazine.	

	Spiral magazine	≡

		Basic solutions
Description of function:	Spiral magazines are roller or slide conveyor devices which are guided around a suitable frame. The workpiece movement is always from top to bottom.	
Drive:	No drive: workpiece movement is effected by force of gravity.	Design examples
Control system:	Generally have no control system.	
Resetting capability	Not generally convertible to other workpiece types.	
		Storage tower (Photo: Liebherr)
Design character- istics:	Storage capacity. Space requirement. Fitting and installation position. Loading and unloading facilities.	
Application:	Are used for loose connection of production devices in gravity linked systems.	
Design forms:	Storage tower	

	Pallet magazine	≡
Description of function:	Pallet magazines are used for storing workpieces in a mainly horizontal arrangement and in a specific orientation. The workpieces in most cases rest in nests, more rarely are clamped, so that when being conveyed the workpiece retains its position. Pallet magazines are in most cases stackable.	**Basic solutions**
Drive:	Pallets generally have no drives of their own. They are transferred by conveyor devices or by gravity.	**Design examples**
Control system:	May be provided with coding.	
Resetting capability	Pallets are generally designed for the workpiece. When using interchangeable workpiece fixtures, they may be adapted to different workpieces within certain limits.	
Design characteristics:	Storage capacity. Resetting capability. Space requirement.	
Application:	Pallets are used primarily in loosely linked production processes.	
Design forms:		

	Tube magazine	≡
Description of function:	Tube magazines are used for storing mainly stackable flat parts, mostly in a vertical arrangement. The workpieces are positioned in most cases by a step device, which is designed as a pusher.	**Basic solutions**
Drive:	Generally no drive: workpiece movement is effected by gravity. In the case of lifting piston magazines, the workpiece movement is effected by means of cylinders.	**Design examples**
Control system:	The segregating device is generally connected to the control system of the production equipment.	
Resetting capability	Not generally convertible to other workpiece types (except possibly for cylindrical parts).	
Design characteristics:	Storage capacity. Resetting capability. Space requirement. Height.	
Application:	Tube magazines are frequently used on presses.	
Design forms:	Bar magazine, lifting piston magazine, zigzag magazine.	

	Belt conveyor	→
Description of function:	Workpieces may be moved on the belt conveyor in cradles or clamped in fixtures.	Basic solutions
Drive:	Drive generally electric, frequently infinitely adjustable. Either central drive or head drive on the guide pulley. Also possible: drive through pulley motor.	Design examples
Control system:	Only required in interlinked systems.	
Resetting capability	Belt conveyors are suitable for transferring all workpiece groups and they do not generally need to be reset.	Lightweight conveyor (Photo: Robert Bosch GmbH)
Design characteristics:	Dimensions of the feed channel. Space requirement. Load capacity. Conveyor speed.	
Application:	The belt conveyor is suitable for transferring practically all workpiece groups.	
Design forms:	Various smooth belts (rubber, fabric, plastic, metal). Sectioned belts for inclined conveyors, steel plate conveyors, wire belt conveyors, rail conveyors, magnetic belt conveyors for inclined or vertical conveying of magnetic workpieces. Curved belt conveyors (also three-dimensional).	Slide belt conveyors (Photo: Steiff)

	Roller conveyor	→
Description of function:	The workpieces are moved by rotating rollers; either under the influence of gravity or by driven rollers.	**Basic solutions**
Drive:	(a) Drive by gravity, roller conveyor erected at a gradient of approx. 5°. (b) Rollers are in most cases driven by an electric motor through a belt.	**Design examples**
Control system:	Roller conveyors may be fitted with a control system.	
Resetting capability	Roller conveyors are not generally resettable.	
Design characteristics:	Dimensions of the feed channel, load capacity. Output.	
Application:	Roller conveyors are suitable for transferring workpieces generally having a minimum size and a flat surface.	Mini roller conveyor (Photo: Transnorm GmbH)
Design forms:	New design: plastic rollers which are driven by flexible shafts, shape of the roller may be adapted to the workpiece. (Onflex system.)	

	Vibratory conveyor	→

Description of function:	Vibratory conveyors move workpieces forwards by means of tiny throwing movements.	Basic solutions
Drive:	The vibrations of the unit are produced either by an eccentric drive or more often by electromagnetic means.	Design examples
Control system:	In linked systems, vibratory conveyors are generally connected to control systems. Speed is generally steplessly adjustable.	
Resetting capability	Vibratory conveyors are operated flexibly as a pure transfer device. All workpiece groups, except spherical parts, may be transferred.	Small vibratory conveyor KF 12-1/60/300 (Photo: AEG-Telefunken)
Design characteristics:	Layout of the feed channel in the area. Dimensions of the feed channel. Space requirement. Feeder speed.	
Application:	Storing and transfer of small and medium-sized workpieces in large numbers, which are mainly fracture- and surface-insensitive.	
Design forms:	Rail channel. Spiral conveyor.	

111

	Walking beam conveyor	→
Description of function:	The workpieces positioned on the support beam are lifted with the upward movement of the lifting beam, conveyed in the transfer direction with the horizontal movement and lowered onto the support beam again with the downward movement of the lifting beam. Vertical and horizontal movements may be superimposed in this case.	Basic solutions
Drive:	The drive may be provided with an electric motor and crank or with pneumatic or hydraulic lifting cylinders with lifting rods.	Design examples
Control system:	The unit is generally used in interlinked systems and is equipped with a control system.	
Resetting capability	Generally not convertible for other workpieces since the workpiece fixtures and the step lengths are designed for one workpiece.	Walking beam feeder (Photo: KK-Automation)
Design characteristics:	Dimensions of the feed channel. Space requirement. Load capacity. Feeder speed. Step/pitch. Length of stroke	
Application:	Cycled transfer and feed into machine tools, partially with storage function.	
Design forms:	Lifting step conveyor	

	Vibratory or bowl feeder	
Description of function:	Vibratory bowl feeders produce micro-projectile movements. These movements are used to feed out the workpieces along a rising spiral on the inside of the bowl.	Basic solutions
Drive:	The vibrations of the unit are produced either by an eccentric drive or more often by an electro-magnetic drive.	Design examples
Control system:	High duty vibratory feeders are equipped with a call-off control system which keeps filling level constant within the limits in the bowl. In linked systems, vibratory feeders are generally connected to control systems.	
Resetting capability	Vibratory feeders are operated flexibly as pure transfer devices. All workpiece groups except spherical parts may be transferred, but these feeders are used mainly for sorting. For this purpose they are equipped with mechanical baffle plates which must be adapted to each workpiece.	Vibratory bowl feeder (Photo: Vibratorentechnik)
Design characteristics:	Capacity of the bowl layout of the feed channel in the working area. Dimensions of the feed channel. Space requirement. Transfer speed.	
Application:	Storage, transfer and sorting of small and medium-sized workpieces in large quantities, which are mainly fracture- and surface-insensitive.	
		Vibratory bowl feeder Sorting bowl KSB-N (Photo: RNA)
Design forms:	Most common form: vibratory bowl (stepped container, cylinder container). Other form: spiral conveyor, high duty vibrating conveyor with separate hopper. Accessories: noise insulating cover	

113

	Elevating feeder conveyors	

		Basic solutions
Description of function:	These conveyors feed workpieces out of a hopper with a revolving belt (or chain) fitted with carriers.	
Drive:	Generally electric drives are used.	Design examples
Control system:	The unit can generally be connected to a control system, particularly in the case of interlinked systems.	
Resetting capability	When a workpiece is changed the carriers must be adapted to the new workpiece. However, it is not possible to adapt the carriers to all workpiece groups.	Elevating hopper feeder (Photo: Henri Manigley)
Design character- istics:	Capacity of the hopper. Conveyor speed. Dimensions of the feed channel Load capacity. Space requirement.	
Application:	Storage, transfer and orientation of workpieces in average to large quantities. Used for small to medium workpiece dimensions.	
Design forms:	Workpiece exit at the side by means of a discharge slide, on which tooling is fitted for orientating the workpieces.	Chain conveyor KF-V-H System Akoplan (Photo: Gehomat)

114

	Rotary disc hopper feeder	
Description of function:	In the rotary disc feeder, workpieces are transferred from store in pockets to a discharge channel, and orientated in the process.	Basic solutions
Drive:	The drive is electric.	Design examples
Control system:	The unit is generally equipped with a control system in linked systems.	
Resetting capability	Because of the form-dependent elements, the unit can only be converted for other workpieces with great difficulty. Limited convertibility may be obtained by replacing the rotary disc.	Rotary disc feeder for orientating rotation symmetrical workpieces (Photo: ASK)
Design characteristics:	Dimensions of the feed channel. Discharge output. Space requirement. Storage capacity.	
Application:	Storage, transfer and orientation of relatively simple workpieces in large quantities.	
Design forms:	Axis of rotation of the rotary disc may be inclined to the perpendicular.	Rotary disc feeder (Photo: Bergmann)

	Rotating drum feeder	⟨symbol⟩

		Basic solutions
Description of function:	In rotating drum feeders workpieces are transferred from a store by centrifugal force, or by form-dependent fitting into a discharge channel orientated in the process.	
Drive:	The drive is generally electric.	Design examples
Control system:	The unit is generally equipped with a control system in linked installations.	
Resetting capability	Because of the form-dependent elements the unit can only be converted for other workpieces with great difficulty.	Rotating drum feeder BTR (Photo: Bergmann)
Design characteristics:	Dimensions of the feed channel. Discharge output. Space requirement. Storage capacity.	
Application:	Bunkering, transfer and sorting of relatively simple workpieces in large quantities.	
Design forms:	Rotary disc feeder. Drum feeder with rotating base. Rotary drum feeder.	Distribution drum (Photo: IWECO)

	Centre board hopper	
Description of function:	In the centre board hopper the workpieces are transferred by linear or circular movement of the centre board out of the hopper to a discharge slide (by gravity), where the parts can be further orientated.	**Basic solution**
Drive:	Generally electric drive, or in the case of linear movements, also pneumatic.	**Design examples**
Control system:	The unit may generally be connected to a control system, particularly in linked installations.	
Resetting capability	Convertible for similar parts by baffle plate replacement on the discharge slide, regardless of the workpiece length.	Centre board hopper Pneumasort (Photo: MFM)
Design characteristics:	Capacity of the hopper. Discharge output. Dimensions of the feed channel. Space requirement. Stroke rate of centre board.	
Application:	Storage, transfer and orientating of workpieces in medium quantities.	
Design forms:	Tilting hopper. Lifting slide.	

117

	Escapement devices	
Description of function:	Escapement devices generally have the task of segregating magazined workpieces (mostly in line) by releasing one particular workpiece and blocking the rest of the workpiece flow.	**Basic solutions**
Drive:	Dependent on the design of the escapement parts.	**Design examples**
Control system:	The control of the escapement is generally provided by the control system of the feeding device in which they are installed.	
Resetting capability	The escapement parts are adapted for a specific workpiece and function, and are therefore very difficult to convert.	
Design characteristics:	Dimensions of the feed channel. Shape of the feed channel. Type of escapement.	
Application:	Escapement devices are used mostly for segregating workpieces from magazines.	
Design forms:	Detent gate, blocking mechanism, paddle wheel, rollers, belts, chain with carrier, slide, screw, disc, plate wheel, drum, gripper.	

118

principles were always described according to the same plan. To illustrate the basic solutions examples already realized are illustrated, where possible.

4.4.2 Examples of flexible feeding devices
By means of several examples the following describes how flexible orientating devices can be provided, by combining conventional feeding devices with sensors, these devices being especially suited for use at workplaces which can be automated with industrial robots.

All the examples were developed within the framework of the research and development project promoted by the BMFT, entitled "New handling systems as technical aids for the working process". [Neue Handhabungssysteme als technische Hilfen für den Arbeitsprozeß.]

4.4.2.1 Programmable workpiece recognition equipment
A feeding device for workpiece recognition has been developed by IBP-Pietzsch GmbH, and linked to an industrial robot (PM 12) (Figure 4.10). The workpieces are fed into a V-shaped conveyor in any position and in this conveyor they automatically assume one of a few stable preferential positions. The workpiece is moved past the sensors in the conveyor and scanned without contact. A logic circuit composes the sensor signals to form a screen image of the workpiece. The screen image recognisable by a photo diode field is exhibited in Figure 4.11.

The workpiece image is compared with previously programmed reference images. If there is a correspondence between the workpiece screen image and a reference image, the workpiece is recognised. If there is no correspondence between the workpiece image and a reference image, the workpiece is segregated. The recognised workpieces are gripped by an industrial robot linked to the recognition device, and deposited in a place corresponding to the recognised position.

4.4.2.2 Feeding devices with incoherent-optical correlator for workpiece recognition
At the Fraunhofer Institute for Production Engineering and Automation (Produktionstechnik und Automatisierung) (IPA), Stuttgart, an incoherent-optical correlator developed by the Institute for Data Processing in Engineering and Biology (Institut für Informationsverarbeitung in Technik und Biologie) (IITB), Karlsruhe, has been linked to an industrial robot (Figure 4.12).

Circular workpieces arranged behind one another in any sequence pass on a

Figure 4.10
Programmable workpiece
recognition unit
linked to
an industrial robot.

119

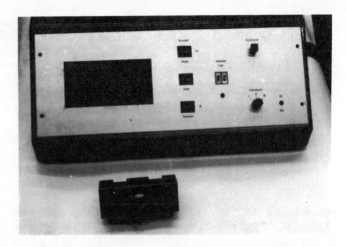

*Figure 4.11
Display panel on
the control desk of
the recognition unit.*

conveyor belt through the field of vision of the correlator their position being measured perpendicular to the direction of conveyance. The output signals of the correlator control the industrial robot so that the gripper moves into the corresponding position and the workpiece is moved into the open gripper by the conveyor belt. By actuating the contact switch in the gripper the latter closes.

The workpiece is then transferred to a preprogrammed position. While the position is being determined, workpieces with incorrect diameters are detected and are not gripped by the industrial robot. The recognition process in the opt.-incoherent correlator takes place by a comparison of the existing pattern marks (workpiece views) with comparable marks in transparent form (so-called masks). The incoherent light beams reflected by the patterns are processed optically and in parallel by simple lens systems. The different workpieces are adapted by changing the comparable pattern in the correlator (4.6).

To evaluate the correlator signals, only a simple electronic decision logic is

*Figure 4.12
Feeding device with
incoherent-optical
correlator.*

Figure 4.13
Television sensor linked
to an industrial robot.

required here. By suitable design of the masks, the workpiece spectrum may be extended to cover any workpieces which exhibit circular workpiece characteristics.

4.4.2.3 Orientation of workpieces by an industrial robot with a television sensor

Mechanical orientating devices are designed for a single workpiece and must be converted at considerable expense when workpieces are changed. Such devices are not suitable, therefore, for small series. To remedy this, optical sensors are linked to an industrial robot to form a flexible orientating device (Figure 4.13).

The television sensor can be programmed simply by presenting the workpiece in all the possible stable positions when positioned against a stop. During the recognition process, it is established which of the programmed positions the workpiece is assuming. At the same time its position at the stop is measured. Both sets of information are transmitted to the control system of the industrial robot, which grips the workpiece and places it in a magazine or machining station.

The workpiece is observed by transmitted light. The black-white transitions are measured and evaluated in a maximum of 5 of a total of 625 television lines. The position of these lines on the screen can be chosen as required (Figure 4.14). The cost of computing time and storage capacity is reduced by this reduction in the signals to be processed, with the result that a microprocessor could be used for the data processing. The recognition process takes approximately 70 ms.

4.4.2.4 Flexible mechanical orientating, feeding and magazining system

Whilst industrial robots are flexible, because of their programmability, and can easily be converted or reprogrammed for different types of workpiece and working processes, devices and magazines offered on the market do not possess this facility.

An elevating conveyor (Figure 4.15) has been developed and constructed at the IPA as an example of a flexible orientating device. The belt inclination, hopper base, discharge and orientating elements of the device are easily converted or reset,

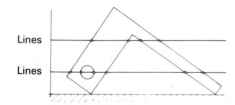

Lines

Lines

Figure 4.14
Line coverage in
the television sensor

121

Figure 4.15
Flexible mechanical
ordering and
feeding system.

Figure 4.16
Loading of foil
magazines with an
industrial robot

and therefore provide the possibility of orientating different workpiece types. By combining with tactile sensors in the discharge slide of the elevating conveyor, which can also be adjusted to different workpiece types, any undefined positions of the workpieces which still remain can be detected. In order to maintain the ordered status of the workpieces for subsequent machining operations once they have been placed in order, simple magazines are being developed at IPA which can be manufactured, for example, at the workplace, for the workpieces to be processed, and can be used as "one-way" magazines. They are manufactured by drawing plastic foil, the workpieces to be magazined being positioned on the plastic foil before drawing and thus simultaneously serving as the deep drawing tool. Figure 4.16 shows such a foil magazine in different loading pallets.

The above-mentioned flexible feeding devices are being tested in an experimental workshop (4.7) together with an industrial robot and a refinement of the television sensor described in Section 4.4.2.2. A drilling station for small batches, available in the industry, was constructed and extensively automated for this purpose in the IPA laboratory. The resetting times for the different workpieces of the workpiece spectrum, which comprises 15 different types, was reduced to a minimum so that automatic machining, even of very small batches, becomes an economic proposition.

5. Application planning

5.1 Introduction

The reasons for the small number of successful applications can be found in both technical and non-technical fields. The main technical obstacles are the absence of sensory capabilities of the handling equipment and the lack of flexibility of the required feeding devices. In the non-technical field, particular mention must be made not only of the commonly found distrust of new technical developments, but also of the lack of knowledge of the possible applications of industrial robots, the uncertainty regarding the cost effectiveness of an application, as well as the lack of suitable aids for implementing application planning.

This chapter is aimed at making a contribution towards removing the last mentioned obstacles. In particular, a guide will be given to show the criteria which must be used to assess the basic applications of an industrial robot, to show how application planning can be rationally implemented, and finally to show how the cost effectiveness of an industrial robot application can be calculated.

5.2 Survey of procedures for methodological application planning

The method of application planning is represented in Figure 5.1 in the form of a flow chart. The starting point for planning is an analysis of the actual condition of the workplaces in the area being examined. This analysis is carried out in two stages. Whilst the first part of the analysis – the "rough analysis" – serves to assess the basic applicability of industrial robots, the second part – the "detailed analysis" – provides for the collection of all the data on the workplaces of relevance to application planning.

On the basis of this detailed analysis, the handling process can be established in the form of a functional plan, in relation to each workplace studied, and performance specifications can be formulated for the handling equipment and feeding devices to be used. In addition modifications are determined which must be carried out on the production units, transfer equipment and controlling devices, in order to achieve full automation of the working process.

Alternative solutions to the proposed operating plans are designed in the next planning stage by the allocation and combination of suitable function carriers. By comparing these alternative solutions in a multi-dimensional evaluation procedure on the basis of cost benefit analysis, the solution which promises the greatest success in technical and economic terms is selected from this range of solutions. Suitable means for the technical realisation of this solution are then

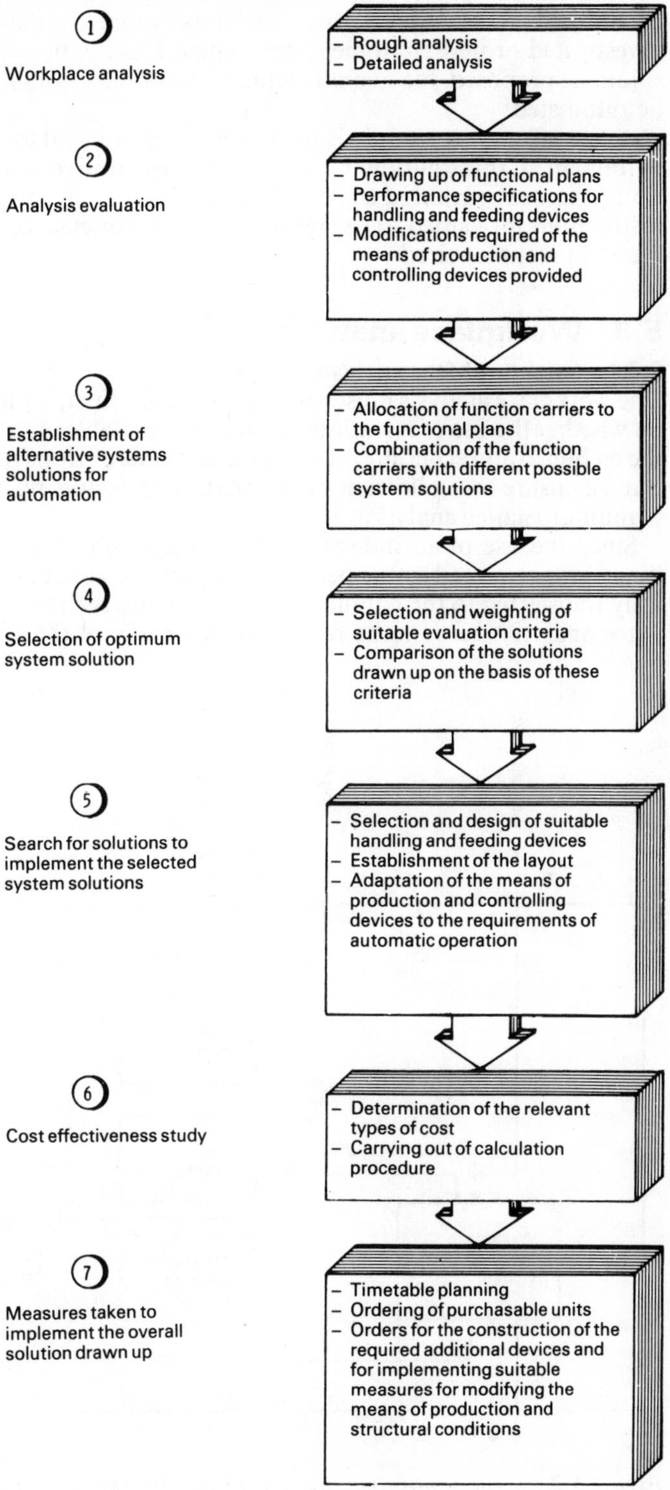

① Workplace analysis
– Rough analysis
– Detailed analysis

② Analysis evaluation
– Drawing up of functional plans
– Performance specifications for handling and feeding devices
– Modifications required of the means of production and controlling devices provided

③ Establishment of alternative systems solutions for automation
– Allocation of function carriers to the functional plans
– Combination of the function carriers with different possible system solutions

④ Selection of optimum system solution
– Selection and weighting of suitable evaluation criteria
– Comparison of the solutions drawn up on the basis of these criteria

⑤ Search for solutions to implement the selected system solutions
– Selection and design of suitable handling and feeding devices
– Establishment of the layout
– Adaptation of the means of production and controlling devices to the requirements of automatic operation

⑥ Cost effectiveness study
– Determination of the relevant types of cost
– Carrying out of calculation procedure

⑦ Measures taken to implement the overall solution drawn up
– Timetable planning
– Ordering of purchasable units
– Orders for the construction of the required additional devices and for implementing suitable measures for modifying the means of production and structural conditions

Figure 5.1
Sequence of application planning of industrial robots.

determined. For this, it is first necessary to establish the layout of the workplace to be automated. Based on this partial solution, available equipment is then investigated or new equipment developed. Finally, by combining the individual solutions proposed, the overall technical solution is obtained for the workplace to be automated.

A cost effectiveness study is then conducted as an aid to deciding whether or not to implement this solution. If the study proves positive, a timetable must be drawn up in the next planning stage for implementing the concept, and the necessary instructions forwarded to the responsible departments for designing and ordering the equipment required.

5.3   Workplace analysis

5.3.1   Structure of the analysis

The workplace analysis must perform two tasks. First, is must provide information on whether the use of an industrial robot is advisable at the given workplace from the technical and economic or from social points of view (rough analysis); secondly it must ensure the collection of all the data on factors relevant to the application planning (detailed analysis) (5.1).

Since the use of an industrial robot is generally only advisable when all the functions performed by the personnel as part of an operating cycle, can be successfully transferred to the handling unit or suitable peripheral equipment, the workplace analysis must not be restricted to a study of the handling and machining

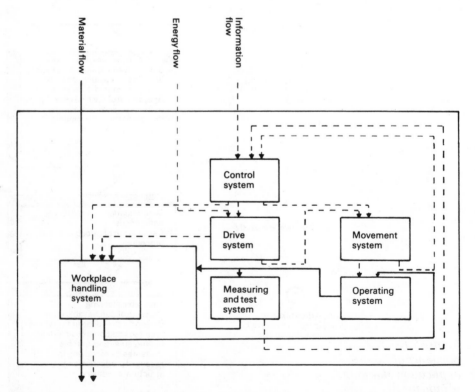

Figure 5.2   Functional structure of a production system (acording to (5.2)).

system, but must also include the recording of the actual conditions of all the sub-systems at the workplace.

The individual subsystems to be studied and their functional relationships are represented in Figure 5.2.

The six subsystems are, in the most developed case, combined to form a single integrated unit. This condition occurs only rarely, however.

Much more frequently:
- the machining system, whose function consists in the alteration of the properties of the workpieces or working materials by direct interaction of tool and material (acting pair),
- the handling system, with which position and location of the workpieces are altered, and
- the control system with which the maintenance of certain conditions or processes is monitored and indicated,

represent independent units which have their own drive, movement and control systems. The reason for this lies in the as yet incomplete automation of the production processes and in the attempts made up till now to increase primarily the output of the machining system.

Another reason for separate units is the fact that the same handling and control systems are often used simultaneously for several machining systems. To permit a quick, and clearly demarcated functional analysis of the activities of the personnel, it is recommended that the analysis be based on this division into three, representing practical conditions. Although this activity or functional analysis is sufficient to provide the basis for a decision on the technical possibility of an industrial robot application, it does not permit an assessment of the economic feasibility of such an application. What is required here is the collation of economic and organisational characteristic data about the given workplace and on the nature of the interlinking with preceding and subsequent workplaces. Furthermore, the working conditions and the possible accident risks must be established within the framework of the analysis. These characteristics provide information on how far automation is required or desirable from a social point of view. Finally, general data required mainly for statistical purposes may also be collected.

If the study of the handling system is further divided into an analysis of the handling process and an analysis of the material to be handled, eight sections are obtained for carrying out the workplace analysis. These sections are represented in Figure 5.3, together with their most important subsections.

5.3.2 Explanation of the analysis
While most of the data contained in the above Figure can be regarded as being uniquely defined, it would appear necessary that for one section an agreement is reached on its contents and evaluation. These terms are explained in the following.

5.3.2.1 Materials to be handled
The materials to be handled cover all items or materials which are supplied to the machining system and discharged from it. Their properties and characteristics have an essential influence on the design of a feeding device. The materials to be handled represented in Figure 5.4a are divided into five groups which may occur in machining processes. In Figure 5.4b, solid parts are further classified according to their group (cf. Chapter 4) on the basis of the physical condition of the goods.

The breakdown according to Figure 5.4 is recommended for determining the nature and quantity of the materials manually handled at the workplace. In

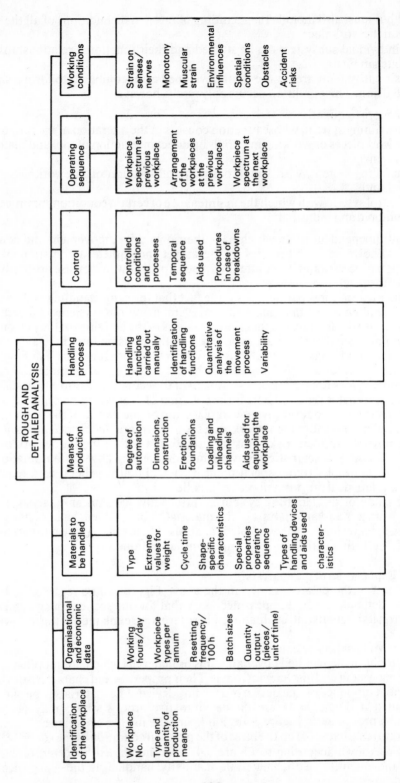

Figure 5.3   Breakdown of rough and detailed analysis (according to (5.1)).

128

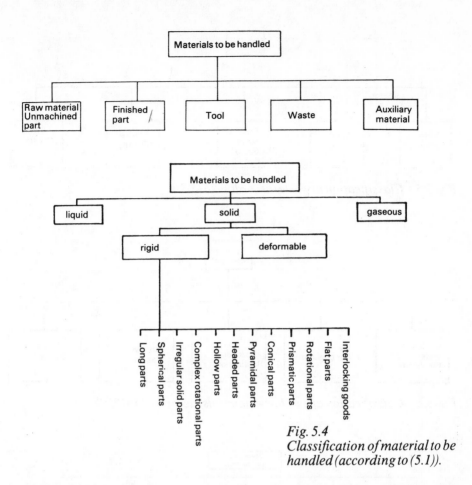

Fig. 5.4
Classification of material to be
handled (according to (5.1)).

addition to the workpieces handled before and after the machining process – which are regarded as unmachined and finished parts respectively for the purpose of the workplace under consideration – tools must also be taken into account if they are changed manually, due to the process involved, or if they are required for manual machining. On the other hand, aids which are used as power amplifiers for extending the human workspace or for protection against injury must not be included, nor tools which are used especially by the human operator for removing waste or supplying auxiliary materials.

Waste and auxiliary materials may be classified as shown in Figures 5.5 and 5.6. By auxiliary materials are meant all those materials which are required for an uninterrupted machining process, but whose alteration or modification is not the purpose of the machining process. In the overwhelming majority of cases these materials are liquid coolants or lubricants. These are regarded as having been manually handled even when their quantity of direction of flow is manually set at least one per cycle.

5.3.2.2   Degree of automation of the process

The degree of automation of the process determines the extent of the modifications required to be carried out on the given means of production to achieve fully automatic operation of the workplace in question, and it must be determined in

129

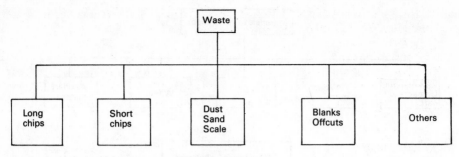

Fig. 5.5   Classification of waste (according to (5.3)).

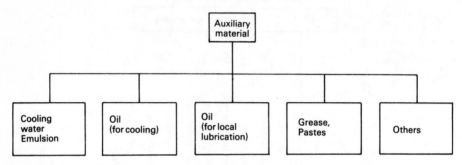

Fig. 5.6   Classification of auxiliary materials (according to (5.3)).

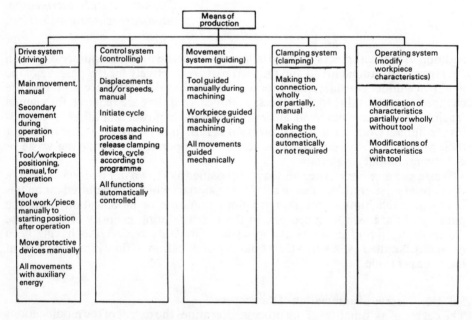

*Fig. 5.7   Criteria for identifying the degree of automation of a means of production
(according to (5.3)).*

130

relation to all the means of production at the workplace. For this purpose, the items listed in Figure 5.7 must all be checked.

5.3.2.3 Handling cycle

The feed functions established in VDI Guideline 3239, described in Chapter 4, are suitable for a qualitative description of the handling process. These functions, however, are not sufficient to obtain exact information on the nature and difficulty of the handling functions to be performed at a workplace.

For this purpose, further specifications regarding the boundary conditions to be observed, and on the frequency and extent of the performance of these functions must be determined in relation to each individual function carried out. When broken down into individual functions, this additional information required is provided as set out below.

Transfer, loading, unloading

These functions may be classified by the following identifying terms:
- path of movement
- speed
- accuracy.

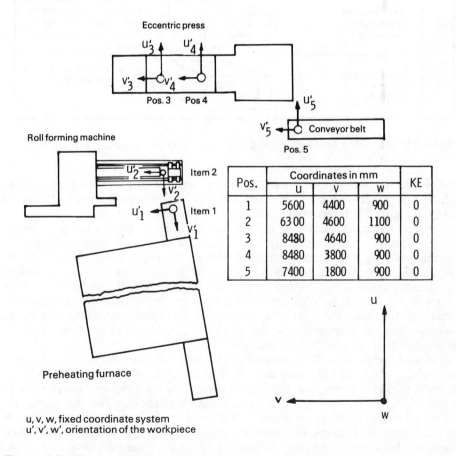

Pos.	Coordinates in mm			KE
	u	v	w	
1	5600	4400	900	0
2	6300	4600	1100	0
3	8480	4640	900	0
4	8480	3800	900	0
5	7400	1800	900	0

u, v, w, fixed coordinate system
u', v', w', orientation of the workpiece

Figure 5.8   Diagrammatic representation of the handling process at a workplace (according to (5.4)).

131

CODE NUMBER KE			
KE	Description	Example	Requirements for kinematics
0	no restriction	——	none
1	approximate straight line	Pos. 1 Pos. 2	Translational axis or simultaneous movements of several axes of rotation
2	exact straight line	Pos. 1 Pos. 2	Translational motion in extension or in parallel with a translational axis
3	Input channel limited on two sides		Translational axis or simultaneous movements of several axes of rotation in the plane of the input channel
4	Input channel limited on four sides		Translational motions in extension of the input channel

Figure 5.9   Meaning and definition of the code number KE (according to (5.4)).

To describe the path of movement, the initial and end points of the path must be recorded, together with possible restrictions. A diagrammatic representation, as shown in Figure 5.8, is recommended for establishing these values.

The representation is based on the machine installation plan of the given work-place. A suitable fixed system of coordinates is chosen for this plan and the coordinate values of the positions to be served determined in relation to this system. The positions are then numbered in the order of the operating cycle, and provided with a code number for describing the possible restrictions of the path of movement shown in Figure 5.9. In addition, a workpiece-related system of coordinates is drawn on the machine installation plan for identification of the workpiece position. This representation is also used for qualitative establishment of the feeding function "turning" described below.

A special case for describing the path of movement is provided when an initial or end point is not fixed but varies from part to part, e.g. in the case of workpieces placed on pallets. In this case it is not the coordinates of one point which must be indicated, but the entire range of coordinates in which the handling points may lie. A further special case occurs when, as on conveyor belts or other continuously operating transfer systems, the handling points are moving relative to the work-place. In this case the extreme values for the speeds must be determined and assigned to each position as the fifth description value.

Futhermore, where the operator handles two or more objects with his hands

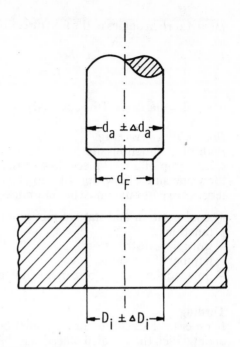

Figure 5.10 Characteristic values for determining the position error $\triangle L$ (according to (5.1)).

simulaneously at the given workplace, the cycle or motions must be determined separately for each and those which take place simultaneously separately identified. Finally it is also necessary to record possible variations of the cycle of motions compared with the one recorded. This information is of importance for subsequent assessment of any convertibility of the handling equipment required.

An exact description of the speed, the second classification characteristic, for the above feeding functions, is only necessary when restrictions are present in relation to this value because of certain properties of the goods to be handled (breakable objects, liquid containers), or because of special conditions of the machining process (painting, arc welding). In this case, the realisable values in each case must be determined separately according to individual functions and operating cycles.

An indication of the maximum permissible positioning error, i.e. the maximum permissible deviation of the actual position from the theoretical one, is used to identify the accuracy with which the above feeding functions must be performed. Whilst this error is determined from the conditions of the particular machining process when using the industrial robot as a processing machine, it is calculated as the smallest possible range of tolerances between workpiece and mating part (Figure 5.10) in the field of workpiece handling, as follows:

$$\triangle L = \pm \frac{(D_i - D_i) - (d_a + \triangle d_a)}{2}$$

$\triangle L$ = positioning error
D_i = inside diameter of the workpiece mating part
d_a = outside diameter of the workpiece

The permissible error may be increased by the use of positioning aids. Such aids include, for example, bevelling, pointing, rounding operations, etc. carried out on the workpiece and/or on the mating part. If it is assumed that the workpiece can be

133

moved at right angles to the feed direction, the positioning error may then be stated as:

$$\Delta L = \pm \frac{(D_i - \Delta D_i) - d_F}{2}$$

where d_F = diameter of the bevelled part.

Branching, converging
Both these functions are subfunctions of the transfer function. They refer to the splitting up of a workpiece flow or the combining of several workpiece flows to form one single flow. In order to classify these functions, the following characteristics in particular must be determined:

– axial position and state of motion of the workpieces before and after the feed function
– workpiece distribution
– number of individual flows
– time sequence of performance of function (continuous, intermittent)

Turning
In order to characterise the feed function "turning" the number of axes of rotation about which the handled object must be rotated and the angles of rotation are the values used.
   These values can be obtained from a diagrammatic representation, as shown in Figure 5.8.

Ordering
Ordering is described quantitatively by the number of undefined rotational and translational degrees of freedom of the objects. Examples of this are given in Figure 5.11.

TRANSLATIONAL LEVEL OF ORDER		ROTATIONAL LEVEL OF ORDER	
Number of undefined degrees of freedom	Explanation	Number of undefined degrees of freedom	Explanation
0	Objects are located at a defined point	0	Position of objects is defined on all axes
1	Objects arranged arbitrarily on straight lines	1	Position of objects arbitrary on one axis
2	Objects arranged arbitrarily in one plane	2	Position of objects arbitrary on two axes
3	Objects arranged arbitrarily in space	3	Position of objects arbitrary on all axes

Figure 5.11   Characterisation of the arrangement of the objects being handled (according to (5.1)).

134

By comparing the orientation of a workpiece at different points in the operating cycle, information is obtained on the extent of the ordering to be carried out. "Translational" ordering is necessary when the coordinates of the object are defined by at least one degree of freedom less at the starting point of a movement section than at the end point. The same applies to rotational ordering.

Metering
Allocation means separating a defined quantity from a larger quantity. Classification characteristics for this are:
– nature of the allocated objects
– quantity and distribution of the objects from which the defined quantity is to be allocated
– allocated quantity

Clamping, releasing
Clamping means securing the position of an object so that it cannot become detached. The following are used for classifying clamping:
– method of making the connection
– clamping force

Binning, magazining
The random and orderly storage of workpieces in containers suitable for them, for forming stocks before and after the production devices, are termed binning and magazining respectively. Both functions do not therefore represent active, but passive processes. They must not be confused with loading and unloading parts into or out of a store. In the workplace analyses these subfunctions are described in greater detail by the nature and number of stores used.

Positioning
An indication of the positioning accuracy, in conjunction with the feed function "loading", makes a more detailed description of "positioning" superfluous.

5.3.2.4   Control functions
In analysing the control functions performed by man, one is faced with the difficulty that many factors, procedures and conditions of processes are performed by man "incidentally", without aids, and are not specified in detail.

As part of the rough analyses, therefore, only those control tasks can be included which
(a) are performed by man with aids
(b) are performed without aids, but are apparent as part of the human operating process and
   – are required to avoid serious consequential damage, or
   – are performed regularly to ensure fault-free processing, or
   – are necessary for the continuation of the process.

This method of analysis is undoubtedly not wholly satisfactory, as it depends on the skill and experience of both the operator and the person entrusted with the analysis, but the most important control tasks can be established by questioning and observing the operator.

5.3.2.5   Working conditions
The working conditions can be evaluated on the basis of the analytical job evaluation for the metal industry in North Württemberg/North Baden (5.5).

Evaluation factor \ Characteristic Basis of evaluation	Strain on senses and nerves	Monotony	Strain on muscles	Noise
	Determining factors are the requirements relating to seeing, hearing, feeling, touching, to perceive the observed condition or process	Monotony is measured by the frequency with which changes take place in the working process, and by the necessity for independent thought and decision	Determining factors are amount, duration and distribution of the load.	Determining factors are volume, nature of the noise (frequency) and and periodic variation of its effect
Very low loading	Observing simple processes	Individual production of larger components or structures, e.g. press tools	Sitting activity, handling of light workpieces	Little or no machine noises, e.g. in the control room
Low loading	Observing complex processes, e.g. on NC (numerically controlled) boring mill or milling machine	Small- or medium-batch production of complicated components with frequent programme changes, e.g. on NC machines	Standing activity, handling of light workpieces	Machine noise without particular noise peaks
Average loading	Performance of accurate, non-routine work, e.g. in tool construction	Medium- to large-series production of simple components with occasional programme changes, e.g. feeding sheet metal into presses	Standing activity with frequent bending down and lifting of parts	Noise of sheet metal presses
High loading	Continuous performance of difficult control tasks on small or high precision parts	Mass production with infrequent changes, e.g. car assembly	Standing activity with regular handling of parts over 10 kg in weight	Noise from forging hammers

Figure 5.12   Examples of evaluation of working conditions (according to (5.1)).

Failing this, it is recommended that the uniformity of the work for the human operator be assessed on the basis of the evaluation factor "Monotony".

The basis for this assessment is the frequency with which changes occur in the work and the necessity for independent thought and decision. For monotony, as for all other working conditions, only a few evaluation stages should be permitted, e.g. the following four:

- Absence of strain or very low strain
- Low strain
- Average strain
- High strain

As the working conditions, apart from one or two exceptions, are not measurable, it is recommended that they be classified on the basis of established examples. Some of these are tabulated in Figure 5.12.

5.3.2.6 Accident risks
As part of the workplace analysis, no objective statements can be made on the possible accident risks. These are only possible on the basis of workplace-related statistics. However, since such statistics are not often available, it would appear desirable to establish possible causes of accidents when conducting the workplace analysis. Attention should be given to the following points in particular:

- risk of becoming caught in the case of direct intervention of the human operator in the working area of the machine
- burning risks due to hot machine parts or workpieces
- risk of being hit by heavy workpieces
- injury risks due to projecting levers or sharp edges
- injury risks from particles or auxiliary materials flying around
- absence of, or inadequate safety devices

5.3.3 Questionnaire
A questionnaire has been drawn up to collate the information described. The questionnaire for the rough analysis is shown on the following pages and is provided with suitable spaces for entering data on quantifiable features. For establishing unquantifiable features selection lists with possible alternatives are drawn up in addition to the individual characteristics. In this case the data is entered by "ticking-off" the appropriate alternative. This "ticking-off method" firstly permits rapid data recording at the workplace, and secondly ensures that the same terms are always used.

5.3.4 Execution and evaluation of workplace analyses

5.3.4.1 Survey
Figure 5.13 provides a survey of the workplace analysis procedure. First the workplace spectrum to be examined in the analysis must be established. It should be extended as far as possible to ensure that the automation measures to be taken are applied first of all where they appear to be particularly urgent from the economic, technical or social viewpoints. Furthermore, when establishing the area of investigation, care should be taken to ensure that a workplace is never considered in isolation but always with those before and after it. For not until such an analysis has been carried out does it become possible to discover changes in planning leading to simplified requirements for handling and thus for the automation equipment. For example, it is recommended that an area limited according to the range of parts or the machining process be fully analysed.

WORKPLACE ANALYSIS

Feature	Alternatives and explanations	Specifications
General data		
1. Workplace no.		. . . .
2. Analysis of firm		. . . .
3. Date of analysis		. . . .
4. Issuer		. . . .
Workplace Identification		
5. Task for industrial robot	workpiece handling only	☐
	Handling with workpiece handling	☐
	Handling with machining	☐
6. Production process	Casting	☐
	Forming	☐
	Parting off	☐
	Joining	☐
	Modifying material properties	☐
	Coating	☐
	Manual work station	☐
Organisational/ economic data		
7. Working hours/day		. . . .
8. Workers/shift		. . . .
9. Workplaces of similar type	in the same firm	. . . .
10. Number of different workpieces per annum		. . . .
11. Batch size (pieces)	on average	. . . .
12. Cycle time (mins.)	on average	. . . .
13. Resetting frequency/ 100 h		. . . .

Sheet 1

138

WORKPLACE ANALYSIS

Feature	Alternatives and explanations	Specifications	
		Blank	Finished part
Material to be handled			
14/15. Number of workpieces of the same type per cycle		. . . .	. . . .
16/17. No. of different workpieces per cycle		. . . .	. . . .
18/19. Max. workpiece weight (kg)	where several workpieces are handled simultaneoulsy: sum of workpiece weights	. . . .	. . . .
20/21. Max. dimensions (mm)		. . . .	. . . .
22/23. Physical condition	rigid	☐	☐
	deformable	☐	☐
	liquid	☐	☐
	gaseous	☐	☐
24/25. Nature of workpiece		. . . .	. . . .
		. . . .	. . . .
		. . . .	. . . .
26/27. Material		. . . .	. . . .
		. . . .	. . . .
28/29. Sensitivity	none	☐	☐
	fragile	☐	☐
	surface sensitive	☐	☐
	impact, shock sensitive	☐	☐
	other	. . . .	. . . .
30/31. Temperature (T)	T = room temperature	☐	☐
	$T- \leqslant 200°C$	☐	☐
	$-200°C < T- \leqslant \ 0°C$	☐	☐
	$0°C < T- \leqslant 400°C$	☐	☐
	$400°C < T- \leqslant 800°C$	☐	☐
	$T- > 800°C$	☐	☐
32/33. Other special properties	no disturbing props.	☐	☐
	wet	☐	☐
	oily, greasy	☐	☐
	chips	☐	☐
	porosity	☐	☐
	other	. . . .	. . . .

Sheet 2

139

WORKPLACE ANALYSIS

Feature	Alternatives and explanations	Specifications	
Material to be handled		Blank	Finished part
34/35. Visual inspection of workpiece	Yes	☐	☐
	No	☐	☐
36/37. Check of dimensions	not required	☐	☐
	manual, with universal measuring device	☐	☐
	manual, with permanently adjusted testing device	☐ ☐	☐ ☐
	semi-automatic, with indication	☐	☐
	fully automatic inspection device	☐	☐
38/39. Additional workpiece inspections		. . . .	
Handling task			
40. Handling function performed	Binning	☐	
	Magazining	☐	
	Placing in order	☐	
	Metering	☐	
	Turning	☐	
	Input, output, transfer	☐	
41. Are simultaneous co-ordinated movements of both hands necessary?	Yes	☐	
	No	☐	
42. Do occasional variations occur in the handling process?	Yes	☐	
	No	☐	
43. Required positioning accuracy		. . . .	

Sheet 3

Feature	
Handling process	
44. Brief description	. . . .
45. Sketch of the handling process and machine installation plan (possibly use photo).	
Example:	

WORKPLACE ANALYSIS

Feature	Alternatives and explanations	Specifications				
Means of production						
46. Number of means of production		. . . .				
		FM1	FM2	FM3	FM4	FM5
47-51. Nature of means of production	Lathe	☐	☐	☐	☐	☐
	Drilling machine	☐	☐	☐	☐	☐
	Milling machine	☐	☐	☐	☐	☐
	Grinding machine	☐	☐	☐	☐	☐
	Swaging and forging machine	☐	☐	☐	☐	☐
	Punching machine and press	☐	☐	☐	☐	☐
	Welding machine	☐	☐	☐	☐	☐
	Furnace	☐	☐	☐	☐	☐
	Cooling device	☐	☐	☐	☐	☐
	Measuring and testing device	☐	☐	☐	☐	☐
	Other	☐	☐	☐	☐	☐
52-56. Residual life (years)		. . . . . .	. . . . .	. . . .	. . . .	. . . .
57-61. Breakdown frequency/h	unknown=99	. . . . . .	. . . . .	. . . .	. . . .	. . . .
62-66. Exact designation of means of production	FM 1 ▷ FM 2 ▷ FM 3 ▷ FM 4 ▷ FM 5 ▷	. . . .				
67-71. Degree of automation of drive	Main movement, maual	☐	☐	☐	☐	☐
	Secondary movement, manual	☐	☐	☐	☐	☐
	Workpiece positioning, manual	☐	☐	☐	☐	☐
	Move safety device	☐	☐	☐	☐	☐
	All drives with auxiliary energy	☐	☐	☐	☐	☐

Sheet 5

142

WORKPLACE ANALYSIS

Feature	Alternatives and explanations	Specifications				
Means of production		FM1	FM2	FM3	FM4	FM5
72-76. Degree of auto-mation of control	Paths of movement and/or speed manually controlled	☐	☐	☐	☐	☐
	Machining sequence manually controlled	☐	☐	☐	☐	☐
	Clamping device released manually	☐	☐	☐	☐	☐
	All functions automatic	☐	☐	☐	☐	☐
77-81. Degree of auto-mation of guidance	Tool guided manually during machining	☐	☐	☐	☐	☐
	Workpiece guided manually during machining	☐	☐	☐	☐	☐
	All movements executed mechanically	☐	☐	☐	☐	☐
82-86. Degree of auto-mation of clamping	Partially or wholly manual	☐	☐	☐	☐	☐
	Fully automatic/ mechanical	☐	☐	☐	☐	☐
87-91. Auxiliary material	No auxiliary material	☐	☐	☐	☐	☐
	Compressed air	☐	☐	☐	☐	☐
	Oil for cooling, lubrication	☐	☐	☐	☐	☐
	Cooling water	☐	☐	☐	☐	☐
	Emulsion	☐	☐	☐	☐	☐
	Pastes, grease	☐	☐	☐	☐	☐
	Rigid bodies	☐	☐	☐	☐	☐
92-96. Feed of auxiliary material in actual condition	Manual, without aids	☐	☐	☐	☐	☐
	Manual with aids, without auxiliary energy	☐	☐	☐	☐	☐
	Manual with aids and auxiliary energy	☐	☐	☐	☐	☐
	Semi-automatic	☐	☐	☐	☐	☐
	Fully automatic	☐	☐	☐	☐	☐
97-101. Frequency of aux. material supply	No data	☐	☐	☐	☐	☐
	1 after several cycles	☐	☐	☐	☐	☐
	1 per cycle	☐	☐	☐	☐	☐
	several times per cycle	☐	☐	☐	☐	☐
	continuously during cycle	☐	☐	☐	☐	☐

Sheet 6

WORKPLACE ANALYSIS

Feature	Alternatives and explanations	Specifications				
Means of production		FM1	FM2	FM3	FM4	FM5
102-106. Waste	No waste	☐	☐	☐	☐	☐
	Long chips	☐	☐	☐	☐	☐
	Short chips	☐	☐	☐	☐	☐
	Dust, scale	☐	☐	☐	☐	☐
	Sheet metal cut offs	☐	☐	☐	☐	☐
107-111. Waste removal in actual condition	Manual, mechanical	☐	☐	☐	☐	☐
	Blowing out, manual	☐	☐	☐	☐	☐
	Manual suction	☐	☐	☐	☐	☐
	Manual flushing out	☐	☐	☐	☐	☐
	Automatic, mechanical	☐	☐	☐	☐	☐
112-116. Frequency of waste removal	No data	☐	☐	☐	☐	☐
	1 after several cycles	☐	☐	☐	☐	☐
	1 per cycle	☐	☐	☐	☐	☐
	Several times per cycle	☐	☐	☐	☐	☐
	Continuous during cycle	☐	☐	☐	☐	☐
117-121. Inspection of operating conditions	Not required continuously	☐	☐	☐	☐	☐
	Required continuously	☐	☐	☐	☐	☐
122-126. Inspection of tools	Not required	☐	☐	☐	☐	☐
	Wear	☐	☐	☐	☐	☐
	Breakage	☐	☐	☐	☐	☐
127-131. Nature of tool inspection	Visual inspection	☐	☐	☐	☐	☐
	With universal measuring device	☐	☐	☐	☐	☐
	With permanently set testing device, manual	☐	☐	☐	☐	☐
	Semi-automatic with indication	☐	☐	☐	☐	☐
	Fully automatic inspection device	☐	☐	☐	☐	☐
132-136. Frequency of tool inspection	Continuous	☐	☐	☐	☐	☐
	Several times per cycle	☐	☐	☐	☐	☐
	Once per cycle	☐	☐	☐	☐	☐
	At longer intervals	☐	☐	☐	☐	☐

WORKPLACE ANALYSIS

Feature	Alternatives and explanations	Specifications			
Means of production					
137. Additional checks		. . . .				
Working conditions		very low	low	average	high
138. Accident risks		☐	☐	☐	☐
139. Monotony		☐	☐	☐	☐
140. Strain on muscles		☐	☐	☐	☐
141. Dirt/dust		☐	☐	☐	☐
142. Oil/grease		☐	☐	☐	☐
143. Temperature		☐	☐	☐	☐
144. Damp, acid, alkali		☐	☐	☐	☐
145. Gas, vapours		☐	☐	☐	☐
146. Noise		☐	☐	☐	☐
147. Vibrations		☐	☐	☐	☐
148. Dazzling/lack of light		☐	☐	☐	☐
149. Risk of catching cold		☐	☐	☐	☐
150. Protective clothing		☐	☐	☐	☐
Machining sequence					
151. Do all workpieces come from the same workplace?	Yes No	☐ ☐			
152. Do other workpieces also run past the workplace?	Yes No	☐ ☐			
153. Do all finished parts go to the same workplace?	Yes No	☐ ☐			

Sheet 8

The workplaces in the established area of study are first subjected to a rough analysis. The object of this rough analysis – as already mentioned – is the selection of workplaces suitable for automation. The selected workplaces are then examined in detail, in order to determine all the factors relevant to further application planning. Performance specifications for industrial robots, peripheral equipment and modification measures at the given workplace may be drawn up as a direct result of this detailed analysis.

5.3.4.2 Results of the rough analysis

The data obtained from the rough analysis provides information on the possibility and urgency of an industrial robot application at a given workplace. To assess both these factors, it is recommended that the procedure represented in Figure 5.14 be followed.

First it should be ascertained whether the workplace in question meets the following three conditions, which must be satisfied when automating handling:

1. The work carried out at the workplace includes handling functions.
2. The same operating processes are repeated several times.
3. The working processes are clearly defined at all times.

It should then be ascertained whether automation is likely to be successful on the basis of the existing technical, economic and organisational conditions. To answer this question, the questionnaires in Figures 5.15 and 5.16 can be used (5.6).

Whilst Figure 5.15 is generally suitable for assessing the prospects of success of handling-specific automation measures, the criteria listed in Figure 5.16 are tailored specifically to industrial robot application. Thus the requirement for the

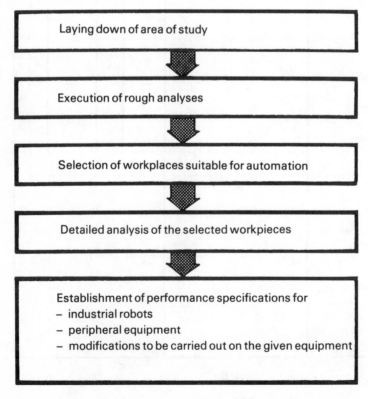

Figure 5.13 Order of workplace analyses

*Figure 5.14
Procedure for
evaluating the
rough analyses.*

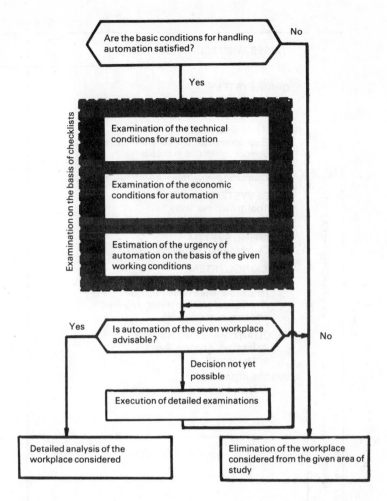

workplace to be operated for at least 2 shifts per day with at least 2 operators per shift arises from economic considerations which were based on the following assumptions:

Investment costs for the automation
(robot and peripheral equipment)   210,000 DM
Wages   50,000 DM/annum
Max. permissible amortization period   3 years

The requirement for limiting the resetting frequency is based on the fact that so far the industrial robot is only slightly faster than the human operator, and therefore the resetting for industrial robots must be of the same order of magnitude as personal delay times, if the same cycle times are to be achieved without the use of additional personnel for setting up as in manual operation. The above-mentioned limit of a maximum of one changeover per shift is calculated on the assumption of a personal delay time of 5% and an average setting up time of 0.5 hour for the handling unit.

The requirement for limiting the workpiece spectrum must be imposed, as at present there are not sufficiently flexible grippers or orientating and magazining devices available. In spite of intensive work in this field it has so far not been

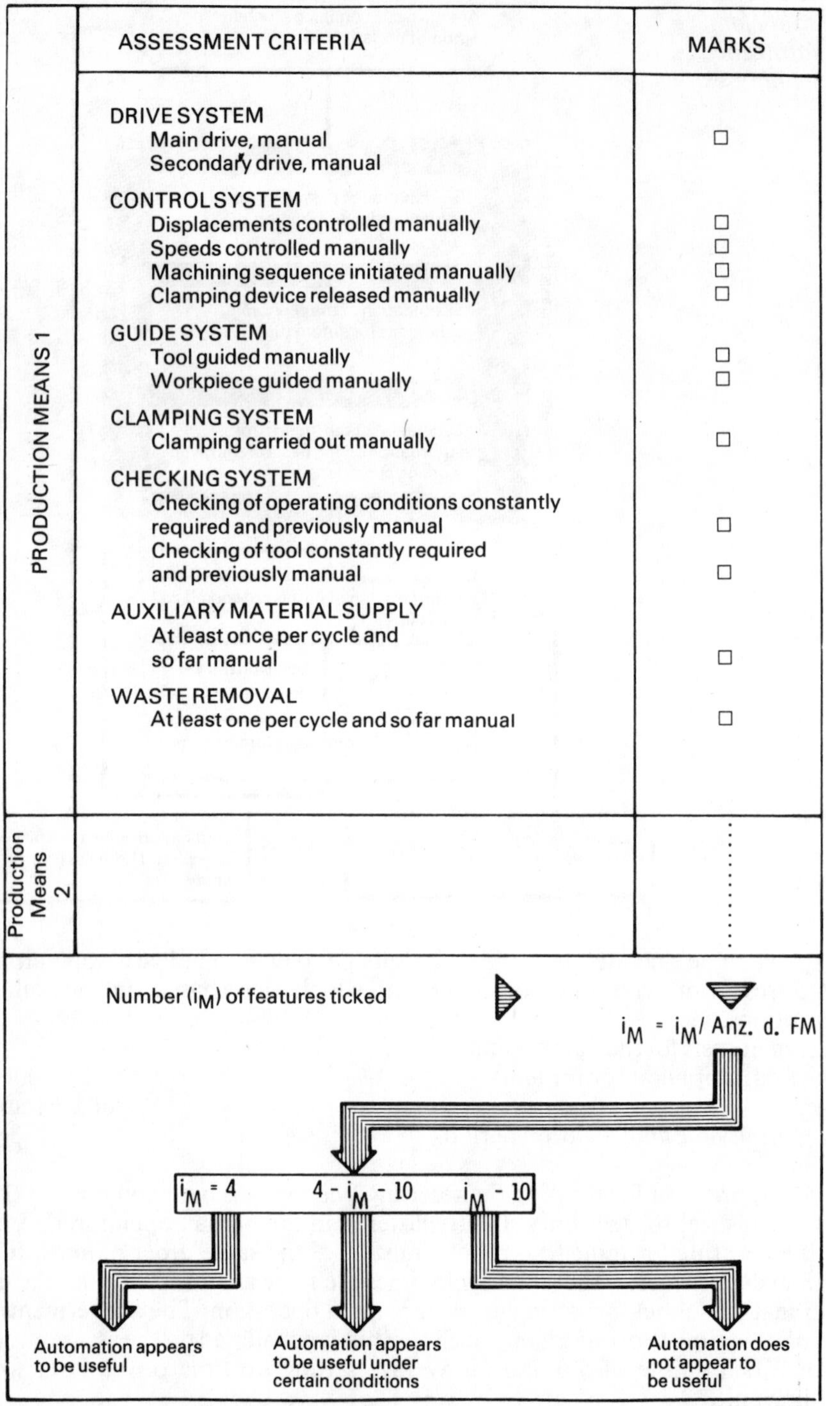

ASSESSMENT CRITERIA	MARKS

PRODUCTION MEANS 1

DRIVE SYSTEM
Main drive, manual
Secondary drive, manual ☐

CONTROL SYSTEM
Displacements controlled manually ☐
Speeds controlled manually ☐
Machining sequence initiated manually ☐
Clamping device released manually ☐

GUIDE SYSTEM
Tool guided manually ☐
Workpiece guided manually ☐

CLAMPING SYSTEM
Clamping carried out manually ☐

CHECKING SYSTEM
Checking of operating conditions constantly
required and previously manual ☐
Checking of tool constantly required
and previously manual ☐

AUXILIARY MATERIAL SUPPLY
At least once per cycle and
so far manual ☐

WASTE REMOVAL
At least one per cycle and so far manual ☐

Production Means 2

Number (i_M) of features ticked

$i_M = i_M /$ Anz. d. FM

| $i_M = 4$ | $4 - i_M - 10$ | $i_M - 10$ |

Automation appears
to be useful

Automation appears
to be useful under
certain conditions

Automation does
not appear to
be useful

Figure 5.15   Technical criteria for estimating the feasibility of handling-specific automation measures.

148

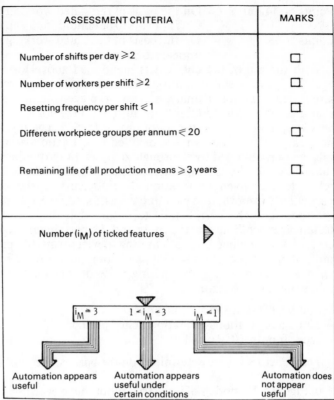

Figure 5.16
Economic criteria
for assessing the
realizability of
handling-specific
automation measures.

ASSESSMENT CRITERIA	MARKS
Number of shifts per day $\geqslant 2$	☐
Number of workers per shift $\geqslant 2$	☐
Resetting frequency per shift $\leqslant 1$	☐
Different workpiece groups per annum $\leqslant 20$	☐
Remaining life of all production means $\geqslant 3$ years	☐

Number (i_M) of ticked features

$i_M \geqslant 3$	$1 < i_M < 3$	$i_M \leqslant 1$

Automation appears useful

Automation appears useful under certain conditions

Automation does not appear useful

possible to find solutions for these three components which allow them to be adapted for a broad workpiece spectrum.

Thus at present the cost of automation is growing considerably with the number of workpieces to be handled. The limiting value of a maximum of 20 different workpiece types assumed in the following table is based on the experience of an abundance of planned applications.

Following the examination of the technical and economic criteria, the extent to which automation should be aimed at must be studied on the basis of the existing working conditions. In order to have as objective a means of comparison as possible for this examination, it is recommended to attach first of all a standardised weighting factor to the individual working conditions featured in the analysis. Figure 5.17 gives a proposed set of these factors for selection which is based on an analytical job evaluation carried out for the engineering industry in North Württemberg/North Baden (5.5).

Based on these weighting factors the total strain to which the operator is subjected at a workplace can now be expressed by a characteristic number when rating numbers are also allocated to each of the four evaluation stages which are used to quantify the individual types of strain. The procedure is illustrated in Figure 5.18 by an example.

For this it is advisable to use the three divisions below following those of the technical and economic criteria:

Points $\leqslant 34$ On the basis of the stated working conditions automation does not appear to be appropriate.

149

Points > 34 but ≤ 68　On the basis of the stated working conditions automation only appears to be conditionally appropriate.

Points > 68　On the basis of the stated working conditions automation appears to be appropriate.

With the aid of the check lists mentioned above for assessing the technical, economic and social conditions for automation, it is often possible to decide already after the rough analysis of a workplace whether to subject it to a detailed examination as to its suitability for automation or not. In some cases the necessarily very simplified check lists cannot provide a clear-cut answer and it then becomes necessary to carry out a more detailed investigation on the basis of conditional information obtained by the rough analysis. In particular it is necessary to check the accuracy of the assumptions made on which the check lists are based on as they relate to the given application. With regard to the economic criteria it is particularly important to examine the assumed investment costs (210,000 DM for the robot and the peripherals) to ensure that they are realistic estimating the technical expenditure for the robot and the peripheral equipment separately. The cost of the feeding and orientating devices can be primarily based on the multiplicity of shapes and the type of components to be handled (see Chapter 4). On the other hand the purchasing price of the robot is chiefly affected by the following required features:

– number of main and auxiliary axes
– number of positions to be approached per axis
– type of control (CP or PTP)

These factors can be determined on the basis of the workpiece and the handling process.

In Figure 5.19 models of industrial robots have been assigned to typical values of these features. This list can serve as a guide for making the above-mentioned assessment.

5.3.4.3　Results of detailed analysis

With the data obtained from the detailed analysis a critical assessment should first of all be made of all the production equipment used at the workplace (production machinery, handling and auxiliary equipment). In this assessment all the modifications should be listed which have to be carried out on the different elements for the

Figure 5.17
Weighting of the strains imposed on the human operator at a workplace.

Weighting factor	Evaluation criteria
1	Vibrations, strain imposed by protective clothing
2	Risk of catching cold, oil/grease, wet/acid/alkali, gases/vapours, dazzling/lack of light
3	Dirt/dust, heat, muscular strain
4	Monotony, noise
5	Accident risks

Evaluation criteria	Wt. factor	Evaluation / Evaluation stages 0=small	1=low	2=average	3=high	No. of points obtained	Max. number of points
Accident risks	5			x		10	15
Monotony	4		x			4	12
Strain on muscles	3			x		6	9
Dirt/dust	3	x				0	9
Oil/grease	2		x			2	6
Temperature	3				x	9	9
Wet/acid/alkali	2	x				0	6
Gases/vapours	2	x				0	6
Noise	4			x		8	12
Vibrations	1			x		2	3
Dazzling/lack of light	2			x		4	6
Risk of catching cold	2	x				0	6
Protective clothing	1				x	3	3
Total						48	102

Figure 5.18 Identification of the strains on a human operator at a workplace (according to (5.7)).

Equipment category		Characteristic features (mean values) max. number of axes	Load capacity kg	Approachable positions per axis	Type of control system	Approximate price DM
A	Very simple units without position control circuits, with a maximum of 2 approachable positions per axis	4	10	2	PTP	50,000
B	Units with or without position control circuits and a limited number of approachable positions per axis	5	15	6	PTP	80,000
C	Units with position control circuits for all axes and any number of approachable positions per axis	6	20	any	PTP	120,000
D	Continuous path controlled equipment	6	15	any	CP	200,000

Figure 5.19 Important characteristics and costs of some industrial robot designs (status: July 1978).

changeover from manual to automated production. These may, among other things, be measures to increase the degrees of automation, to redesign the tools or modify the supply of auxiliary materials or waste removal.

All the workplace characteristics which are independent of the given layout but which influence the selection of the handling systems and peripheral equipment to be used must them be collated in a performance specification. In particular, these are the characteristics listed in Figure 5.20.

151

Workpiece data	Number of workpieces handled per cycle Weight Dimensions State of aggregation Component group Material Sensitivity Temperature
Tool data	Positioning error Resetting time
Handling process	Number of position to be approached Qualitative data (continuous path, point to point)
Working conditions	Heat Dirt, dust Gases, vapours Explosion risk

Figure 5.20
Factors independent of the layout for selecting suitable handling units and peripheral equipment.

As planning continues, this specification serves to formulate equipment-specific requirement profiles, which makes it possible to pre-select suitable equipment types even at this stage.

Furthermore, the handling process can be described functionally on the basis of the results of the workplace analysis. For this it is recommended that the handling process be split up by means of feed functions as set out in VDI guideline 3239, into sub-functions, and represented with the appropriate symbols in functional plans.

Figure 5.21 shows an example of this procedure. The functional description of the handling process provides an excellent basis for designing the handling systems as part of the next planning stage.

5.4 Planning the automated workplace

5.4.1 Formulation and evaluation of alternative system solutions

The separation of possible solutions for automating handling should not be limited to the use of industrial robots but should include an investigation of all technical possibilities. Only in this way can it be ensured that the optimum solution from the technical and economic viewpoints is actually found for a given workplace.

Such a comprehensive method for solving handling problems is presented below. Figure 5.22 provides a survey of this method. An examination of the workplace with regard to:

- the given machine layout
- the number and arrangement of the production equipment and
- interlinking with workplaces before and after the workplace being examined forms the starting point

The object of this examination is to find modification measures which result in simplified requirements for the automation equipment to be used. Figure 5.23 gives a summary of measures frequently offered in this connection.

The modification possibilities found must then be incorporated in the description of the actual condition of the workplace to be automated. In particular, the functional representation of the handling process must be suitably modified, since this forms the basis of further procedure. In the next stage suitable basic solutions and function carriers are sought for the automation in relation to the

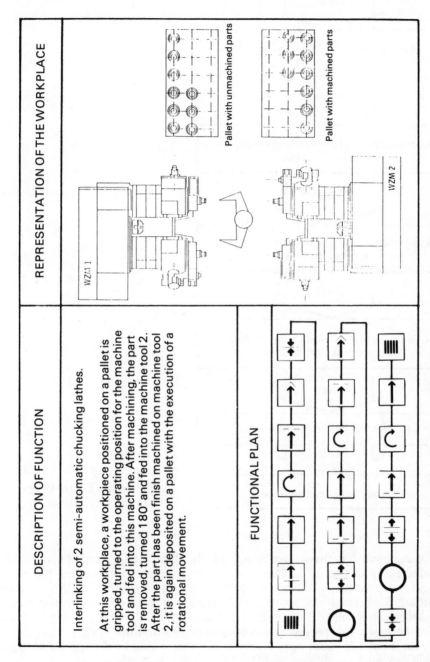

DESCRIPTION OF FUNCTION

Interlinking of 2 semi-automatic chucking lathes.

At this workplace, a workpiece positioned on a pallet is gripped, turned to the operating position for the machine tool and fed into this machine. After machining, the part is removed, turned 180° and fed into the machine tool 2. After the part has been finish machined on machine tool 2, it is again deposited on a pallet with the execution of a rotational movement.

FUNCTIONAL PLAN

REPRESENTATION OF THE WORKPLACE

Pallet with unmachined parts

Pallet with machined parts

WZM 1

WZM 2

Figure 5.21 Example of representation of the handling process in functional plans.

153

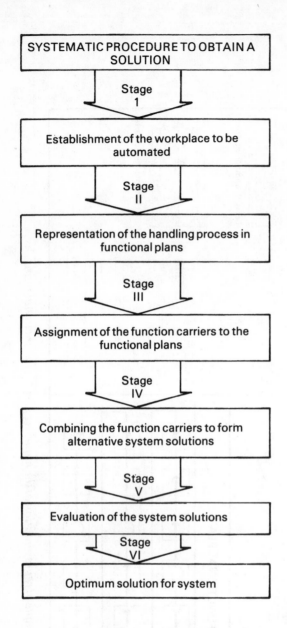

SYSTEMATIC PROCEDURE TO OBTAIN A SOLUTION

Stage 1

Establishment of the workplace to be automated

Stage II

Representation of the handling process in functional plans

Stage III

Assignment of the function carriers to the functional plans

Stage IV

Combining the function carriers to form alternative system solutions

Stage V

Evaluation of the system solutions

Stage VI

Optimum solution for system

Figure 5.22
Procedure for working
out system solutions.

individual sub-functions mentioned in the functional plan. This search is conducted by means of function-specific requirements imposed on the basis of layout-independent workplace characteristics (Figure 5.20). Catalogues offering solutions for orientation, transfer and magazining with devices available on the market can be of considerable help in the search (5.8, 5.9).

Furthermore, this partial task is considerably simplified if right from the start only an industrial robot application is possible owing to the special workplace conditions. Such conditions could be:

– different operating sequences because of certain workpiece characteristics (e.g. segregation of scrapped parts, turning of parts in incorrect position, etc.),

154

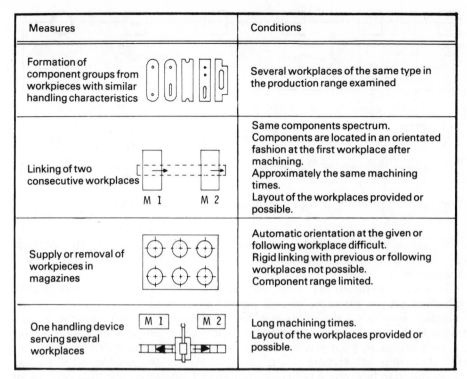

Measures		Conditions
Formation of component groups from workpieces with similar handling characteristics		Several workplaces of the same type in the production range examined
Linking of two consecutive workplaces	M 1 M 2	Same components spectrum. Components are located in an orientated fashion at the first workplace after machining. Approximately the same machining times. Layout of the workplaces provided or possible.
Supply or removal of workpieces in magazines		Automatic orientation at the given or following workplace difficult. Rigid linking with previous or following workplaces not possible. Component range limited.
One handling device serving several workplaces	M 1 M 2	Long machining times. Layout of the workplaces provided or possible.

Figure 5.23 *Technical-organisational measures for automating workpiece handling (according to (5.6)).*

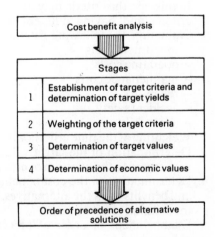

Figure 5.24
Stages in the value analysis.

- different operating sequences due to special operating conditions (e.g. operation during breakdown or cleaning of a means of production),
- changes in the movement sequence after setting the workplace to a different workpiece,
- alteration of the workpiece loading points and the loading and unloading positions after a workpiece change.

In this case it is sufficient to look for suitable basic solutions in relation to those

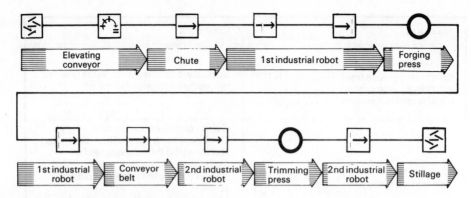

Figure 5.25   Example of assignment of system solutions to functional plans.

sub-functions which cannot be performed by the industrial robot. The solutions selected are then combined to form alternative solutions taking into account the compatibility conditions.

The system solutions are subjected in the next stage to a multi-dimensional evaluation on the principle of cost benefit analysis (5.10). The object of this evaluation is to select the solution whose implementation promises the greatest success both in technical and economic terms. Cost benefit analysis has proved generally satisfactory for problems relating to solution variants, since on the one hand it permits the setting of a number of targets and on the other enables consideration to be given to both quantitative and qualitative characteristics of the alternatives. In particular, it incorporates the following stages (Figure 5.24).

1. Listing of target criteria on the basis of which the alternatives are to be assessed. In this case the criteria may incorporate both quantifiable and non-quantifiable criteria and having different dimensions (e.g. KW, DM). For the present evaluation, the following criteria, among others, present themselves:
   – investment costs
   – operating costs
   – maintenance costs
   – reliability
   – accident safety
   – adaptability
2. Weighting of the time criteria with regard to their influence on the total benefit.
3. Determination of the target values, i.e. evaluation of the alternative solutions with regard to the question of how far the individual target criteria are satisfied.
4. Determination of the cost benefit, i.e. determination of the contributory costs and benefits of all the alternatives. The sum of these provides the cost benefit for the solution alternatives.
5. Establishment of order of precedence for the alternatives using the values obtained.

What results from this is a system solution optimised according to the pre-determined criteria. For a clear representation of this solution the entry of the selected equipment in the appropriate feed function plan is recommended (Figure 5.25).

5.4.2   Solutions for subsystems
In the next stage of application planning, which is the most time-consuming stage,

it is necessary to search for suitable means of implementation for the designed system solution. In this case the following problems in particular must be solved:

- layout planning
- selection of suitable handling equipment
- selection of suitable feeding devices
- automation of operation and control functions
- automation of auxiliary material supply
- automation of waste removal

There exists a strong interdependence between the solutions to some of the subsystems. For example the selection of handling and feeding devices is greatly influenced by the layout. Similarly changes in the layout are often affected by the solutions for the waste disposal and auxiliary material supply systems. Other interrelations exist in cases where the manipulators and feeding devices are used not only for component handling but also for waste disposal and auxiliary material supply. As a result the individual subsystems must be treated in a fixed time sequence which cannot be laid down as a general principle but which must be determined for each application based on the special conditions of the example being considered. It depends very largely on whether new production equipment will have to be acquired for the automated workplace, the robot is to perform auxiliary operations in addition to the handling of the workpieces, and the measures to be taken on the given equipment (waste removal auxiliary material supply, checking machining) influence the layout planning and the choice of equipment. However, the general rule is that the layout planning must be completed before the selection of the equipment and that all planning stages which can affect the geometric relationships of the workplace must be considered before then.

The subsystems which have to be dealt with individually are described below in detail and general procedures are introduced in order – as far as possible – to effect their solution.

5.4.2.1 Problems met when automating machining and inspection functions
Subsequent automation of existing means of production is generally technically possible, but not often economically advisable. Thus where the degree of automation is insufficient, it must first be ascertained, on the basis of the cost of automation and the costs of purchasing suitable production equipment, whether subsequent automation or new acquisition is more economical. This may also lead to the conclusion that the workplace under consideration cannot be economically automated.

The same considerations apply to the automation of inspection, but here the possibility of displacement must also be investigated. The simplest solution to this is represented by the proportional employment of personnel for monitoring all the conditions or processes at several workplaces.

5.4.2.2 Supply of auxiliary materials
In the case of auxiliary material supply, the following parameters are variable, in addition to the type of auxiliary material:

- point in time at which the auxiliary material is supplied in the cycle,
- location of auxiliary material supply,
- quantity supplied.

157

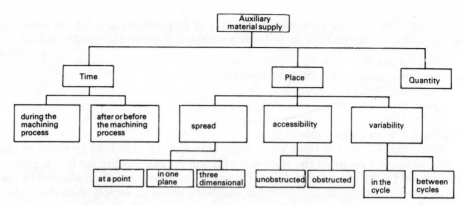

Figure 5.26   Parameters for identifying the technical solution to auxiliary material supply (according to (5.3)).

These parameters can be further subdivided, as shown in Figure 5.26. They represent a decision aid in the search for technical aids suitable for automating the auxiliary material supply, and of particular interest here is the question of whether the auxiliary materials are to be supplied by fixed mechanical devices or by means of an industrial robot. The following principles may be laid down for this:

1. If auxiliary materials have to be supplied during the operating process, fixed mechanical devices should be used.
   Example: Coolant supply for turning.
2. If the point to which the auxiliary materials must be supplied is fixed, or if it is only changed when setting up, rigid and thus cheap special devices can be used.
   Example: Lubricating oil supply to a jig bush.
3. If auxiliary materials have to be supplied to several awkwardly located points between two machining operations, requiring a moving device, the use of an industrial robot is advisable.
   Example: spraying of moulds on pressure casting machines.

5.4.2.3   Waste removal
Whilst the removal of liquids or gases are generally problems which can easily be solved, the difficulty in removing solid waste depends on the quantity, distribution, shape and arrangement of this waste.

In order to automate the removal of solid waste, the following basic measures can be taken.
Active measures:
– mechanical removal
– blowing
– suction
– flushing
Passive measures:
– removal of the waste produced from the machine working area by means of gravity.

Among the active measures it must also be decided whether they can be taken using rigid or moving devices.

For subsequent installation on machines or devices already being used in production, generally only the first three active measures mentioned can be considered. Flushing or passive waste removal must be planned as early as the

158

Factor	mechanical device	suction — with fixed device	suction — with moving device	blowing — with fixed nozzles	blowing — with moving nozzles
Waste quantity	no restriction	no restriction		only small quantities (due to secondary contamination)	
Weight, dimensions of a particle	no restriction	small particles only (max. suction force is 1 bar)		only small particles (depending on available air pressure)	
Wetting with coolants	no limit	for small particle sizes (e.g. grinding dust) no longer usable	Reduction in reliability of waste removal due to increased adhesion forces	reduced range	
Efficiency of waste removal	only when waste is produced at accessible, closely confined points	unreliable	only reliable in the area covered; also at blind holes or inaccessible places	locally good, even in places difficult to reach	good at a point, unreliable for larger areas
Waste removal during machining	hardly possible (collision risk)	possible	hardly possible, collision risk	possible	possible (collision risk lower due to larger than that for suction)
Costs	high	low	high	low	high
Example	Wiper, industrial robot with gripper, brush, etc.	Suction equipment on grinding machines	Suction opening guided by IR (vacuum cleaner)	Fixed nozzle	Compressed air guided by robot nozzle

Figure 5.27   Demarcation of the solutions that can be used subsequently for waste removal from the machine shop (according to (5.3)).

design stage of the machine or plant. The most important characteristics of these possible solutions are summarised in Figure 5.27.

In order to decide whether an industrial robot must be used for waste removal, the same parameters and conclusions as for the supply of auxiliary materials apply.

5.4.2.4 Planning of machine installation

5.4.2.4.1 Factors to be considered

To ensure smooth operation of a workplace automated with industrial robots spatial and material flow boundary conditions must be taken into consideration in planning. In addition to obvious restrictions such as the size of the area available, and structural obstacles difficult to modify such as columns, cable shafts or chip ducts, the position of the conveying routes must also be considered. The supply and removal of unmachined and finished parts should, if possible, be able to take place unimpaired by the erection of production equipment and installation of the industrial robot. The same applies to accessibility of the machines for setting, maintenance, repairs, installation and removal. The areas provided for this must be dimensioned not only on the basis of ergonomic considerations but also according to the aids required during these activities.

Particular importance is attached to the safety problems which arise in the use of industrial robots. These arise when a person enters the workspace of an industrial robot deliberately or accidentally. This risk can be reduced by favourable machine installation and the following points in particular must be borne in mind:

1. Resetting
In workplaces where several machines are to be operated, not all of which are set at the same time, the safety of personnel during the setting of a machine must be guaranteed when all the other machines continue to be loaded and unloaded automatically. The following must be investigated as possible solutions to this:

- Automatic loading and unloading of the machine from a side facing away from the operating side, so that access to the machine lies outside the working space of the industrial robot.
- Favourable installation of the industrial robot to permit simple protection of the workplace.
  Example: Overhead installation and vertical loading and unloading.
- Switching off of handling device during the setting of a machine.

If setting is infrequent and takes only a short time, switching off the entire plant may be a suitable solution. The decision to so this is only possible on the basis of a calculation of the costs.

2. Persons working continuously or temporarily at the workplace
The presence of an operator may be required continuously or at regular intervals for sorting, inspection, delivery and removal of parts. This operator must be guaranteed safe access to the workplace, for which the same criteria as given for setting must be applied.

5.4.2.4.2 Alternative methods of installation

Basically, the following installation systems are available for selection for the installation of an industrial robot:

- mounted on the production equipment
- overhead installation
- floor installation

Mounting on production equipment considerably restricts the area of movement of the handling unit and is therefore generally only usable for 1-machine operation.

In addition, it can only be carried out if the forces and dynamic loads produced by the industrial robot are small in comparison with the forces produced in the production process, and conversely when no sudden loads are transmitted from the production process to the handling equipment.

Consequently, the selection of this installation position must be considered to be an exceptional case within the framework of layout planning. By contrast, over-head and floor installations must be regarded as equivalent alternatives. Their possible applications cannot be defined and distinguished in general terms but must be examined individually in relation to a concrete application. For this purpose, the following criteria must be examined in detail:

Space availability

The available space in front of, beside and above the production equipment frequently determines the position of the installation. For example, if an obstacle difficult to modify (e.g. a crane) is located in the area above the production equipment, only a floor installation can be considered.

Accessibility

Accessibility to the production equipment must be retained for maintenance, repair and setting work. In particular, care must be taken to ensure that the use of special aids which are required for carrying out the above work, e.g. the use of a fork lift truck or industrial truck for resetting, is not obstructed by the installation of the handling unit.

Failure of the handling equipment

A common requirement is that manual operation of the production equipment must be possible, in the event of a breakdown of the handling unit, without the unit having to be removed from the workplace. In this case suitable free areas must be provided in layout planning for the operating personnel. This is frequently only possible by providing overhead installation.

Input and output channels

The position of the input and output ducts on the production equipment also effects the choice of installation position. If the production equipment in a work-place is fitted with vertical channels, overhead installation of the handling equipment is more suitable since the feeding of the production equipment would only be possible with floor installation by means of complicated movements.

The opposite is true for production equipment with only horizontal input and output channels.

Safety precautions

As already mentioned, protection of personnel from the movements of a handling unit is an essential precondition for the use of an industrial robot. In the case of floor installation, a barrier round the entire workspace of the handling unit, using a fixed protective mesh or enclosure is required for this purpose. These safety devices can often obstruct repair work on the occasion of equipment breakdowns and in resetting work. On the other hand, when choosing an overhead installation protection of the input and output channels of the production equipment is generally sufficient.

	FLOOR INSTALLATION	OVERHEAD INSTALLATION
C O N D I T I O N S	sufficient space available in front of and beside the production equipment no vertical input or output channels in the production equipment sufficient floor loading capacity	sufficient space available above the production equipment (no hoists, suction devices, etc.). no horizontal input or output channels in the production equipment sufficient ceiling load capacity
A D V A N T A G E S	simple realisation costs low installation costs simple maintenance and repair of the handling equipment	good accessibility to the production equipment manual operation possible in the event of failure of the handling equipment protection of the input and output channels sufficient on the production equipment
D I S A D V A N T A G E S	accessibility to the production equipment limited, maual operation only possible under certain conditions in the event of failure of the handling equipment protection of the entire working area of the handling equipment required	realisation requires special gantry high installation costs more difficult repair and maintenance of the handling equipment

Figure 5.28   A comparison of floor and overhead installation of the industrial robot.

Construction conditions
It must be ascertained whether both floor and overhead installation of the handling equipment is possible on the basis of the permissible floor and celing loads. Since conversion work to increase the load capacity of individual building elements (reinforcement of ceiling supports, foundations etc.) is generally very expensive, the handling equipment should generally be installed in the position that can be achieved on the basis of the existing structural conditions.

162

Costs

Overhead installation of the industrial robot requires special, often very expensive gantries, and is therefore generally much more expensive than floor installation. In addition to these increased once-for-all costs, suspended installation gives rise to higher running costs due to more difficult repair and maintenance work on the equipment. Thus where no clear advantages have been shown for a particular installation position, floor installation is preferable for cost reasons on the basis of the above criteria.

Figure 5.28 provides a summary of the conditions which must be present to achieve floor and overhead installation, together with a comparison of the advantages and disadvantages of both these possible solutions.

5.4.2.4.3 Alternative installation principles

By a detailed comparison of different possible installation forms, it can be shown that the layout pinciples, shown below in Figure 5.29, are particularly favourable for the layout of a workplace to be automated with industrial robots. In particular, these installation principles show the following important advantages over other solutions:

- minimum number of axes required
- short distances between the individual elements to be served
- small space requirement

Therefore it should first be ascertained in layout planning whether the existing workplace conditions would permit the use of one of the types of installations shown. If so, the following basic principles should be taken into account:

- The arrangement of the production equipment on a circle requires much less space than installation in a line or double line. The feasibility of this form of installation must therefore be examined first. Only if not all the production equipment can be aranged in a circle, or if the radius of the circle is so great that operation with conventional equipment is no longer possible, should the equipment be installed in parallel.
- Installation in concentric circles is only possible when the handling unit is

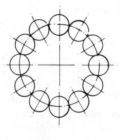

Radially in a circle

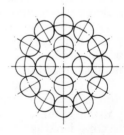

Radially in concentric circles

*Figure 5.29
Favourable
installation
principles from
the handling
viewpoint*

Parallel in line

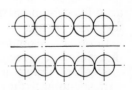

Parallel in double line

163

installed overhead, or if there are considerable differences in the height of the production equipment to be operated.

- For handling processes in which two or more defined linear movements following directly one after the other are required, a unit with a spherical working space is not advisable. For this purpose, movements having at least two axes must be jointly controlled (continuous path control system). In addition, position correction in the gripper is necessary and it is not possible to produce parallel movements in two different planes with the basic axes only.
- For movements which require three defined translational directions, industrial robots with a rectilinear workspace are a suitable solution. It must be remembered, however, that altogether only three defined directions are possible with this construction without continuous path control.

5.4.2.4.4 Procedure for layout planning

Particular importance must be attached to systematic and careful layout planning as part of application planning, as the results of this planning stage not only have a considerable influence on the type and design of the handling unit, but also determine important factors such as:

- safety
- accessibility for maintenance and setting
- output
- space requirement and
- conversion expenditure

In planning, simultaneous optimisation of all the factors influenced by the layout is not possible as some of these contradict each other. Thus a reduction in the space requirement generally reduces accessibility to the production equipment. The same is true for a machine layout which imposes only minimal requirements on the kinematics of the handling equipment which can generally only be achieved by modifying all the production units. Because of these contradictions and the fact that individual solutions elude an objective evaluation (accessibility, safety), it would not appear possible to find a universally applicable procedure which necessarily leads to the optimum layout. In this case, a determination of the optimum solution would only seem feasible by comparing different alternative solutions. As there is generally an abundance of layout variants available, and as their comparison is very involved, suitable computer programs were developed at an early stage for performing this task (5.4, 5.11, 5.12). A simplified method will be presented below which produces the planning solution relatively quickly, even when using a manual procedure. Figure 5.30 gives a survey of this method.

The given machine installation forms the starting point for the layout planning. It must first be ascertained whether automation with equipment on the market is possible in this installation. In this case the possibility of floor installation should be considered as a priority, for the reasons mentioned previously. Equipment with three translational basic axes can only in exceptional cases be used at existing workplaces because of their very limited space. However, fairly high chances of success are offered for industrial robot types with spherical and cylindrical workspaces and for those with articulated joints.

In order to examine the application of this equipment in an individual case a diagrammatic procedure is recommended, as illustrated in Figure 5.31. In this case circles are drawn around the positions to be approached, with the radius corresponding to the maximum range of purchasable equipment in direction x.

If these circles do not have a common point of intersection, the positions lie too

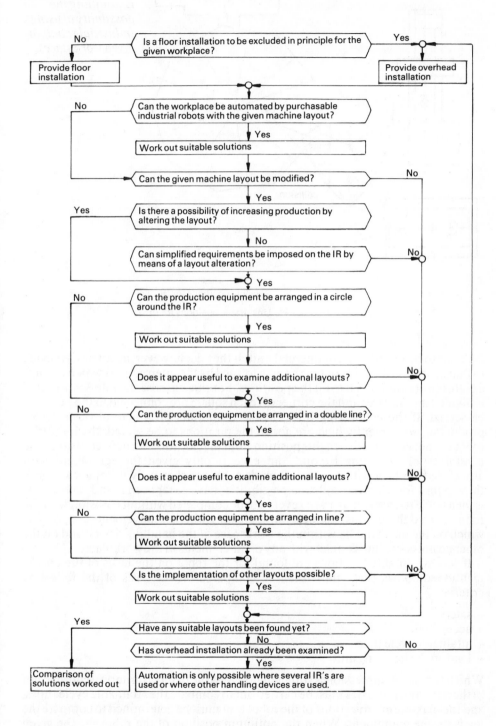

Figure 5.30   Procedure for layout planning for an industrial robot application.

165

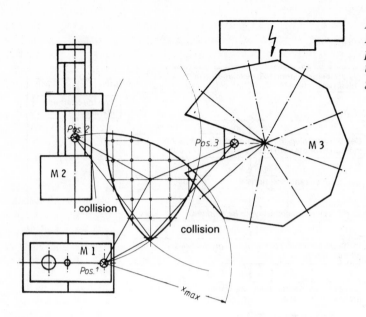

Figure 5.31:
Examining the
possibility of using
industrial robots at
given workplaces.

+ Possible positions for IR
+ Unsuitable positions for IR

far apart to be served by a commercial unit. If they do, however, an attempt is made to find a location for the robot in the area formed by the intersections which ensures a collision-free approach to all the positions. For this purpose the intersection area is divided into squares and the corners of each square are connected to the points to be served. If the connecting lines intersect contours of structural obstacles or production equipment which are above the positions to be served, the industrial robot cannot generally reach this position from that location without collision. The possibility of "reaching beyond" an edge is only given for equipment with articulated joints or with a spherical workspace, and then only in some cases. For those points for which no collision possibilities have so far been found, a check list is then made to ensure that a workcycle can be completed without collision with the robot located there. This is done by investigating whether movements are possible which on the one hand lie within the workspace of the handling device and on the other do not come into contact with any of the elements in the workplace.

If several suitable locations are found for the robot on the basis of the checks mentioned above, these must then be compared on the basis of the following criteria:

– safety
– accessibility
– attainable cycle time
– load on the handling unit

Whilst the visual impression is sufficient to assess the first two criteria a comparison of the handling processes must be carried out to compare the attainable cycle times, and the maximum projections of the robot arm must be determined to compare the loads on the equipment. When the optimum position of the robot for the given machine installation has been determined in this way, it must be ascertained in the

166

next planning stage whether a modification of the given layout is possible and reasonable. In this case the required conversion expenditure, the given room conditions, any increases in output and possible simplifications for the automation equipment must be used as assessment criteria. If these criteria favour a layout modification, the possibilities of a machine arrangement in a circle, in a double line and in a single line must be examined in succession. When examining suitable layouts an attempt should first be made to find solutions in which the conversion expenditure is as low as possible in relation to the given machines. Another important criterion when working out layouts is the space requirement. Furthermore, consideration must be given to the criteria listed in connection with the choice of the robot location described above.

Finally the basic solutions worked out, together with any other layout variants, must be compared. In carrying out this comparison the economic analysis procedure is again recommended, if the decision for one particular solution is not obvious because of the small number of possible solutions or because of a clear difference in characteristics. In this case the above-mentioned criteria must again be used for the evaluation. If the checks described do not indicate that automation is possible with purchasable handling equipment for a given workplace, an attempt must be made to achieve automation by the use of several units or the use of different handling devices. Following this the construction of a newly designed handling unit tailored to the particular case must also be considered. However, this solution is generally only feasible when the equipment can be constructed, largely by means of available parts or modules. Suitable aids and procedures were developed for solving this problem in (5.4 and 5.13).

5.4.2.4.5 Examination of the layout principle

To compare the different possible layout principles it is sufficient to subject the travel distances and times for the handling equipment to a qualitative analysis. Following that, however, it is necessary to determine the movement time more accurately for the selected layout in order to obtain information on the output rates for the workplace in the automated condition. Finally, a decision as to whether the layout selected can be realised can only be taken on the basis of the calculated output rates.

The travel times for the industrial robot depends largely on the gripper equipment selected and for this reason they must be determined successively for the following basic gripper types (cf. Chapter 2.4):

– simple gripper
– double gripper
– 2-arm type

The calculated values later form an important decision criterion for selecting the gripper design. The travel time is determined in each case for the motion which allows the working cycle to be completed in the minimum time. It is generally possible to establish this "minimum time" distance by means of the layout plan without any problems; only in individual cases is a direct calculation not possible, e.g. if there is an obstacle in the travel which can be bypassed by different methods (swivelling over the obstacle; swivelling past with arm retracted). In this case suitable travel times must be determined for both alternatives, and an approximate value for the time may be obtained from the following formula:

$$t_v = \Sigma (\frac{s}{v_{max}} + t_1 + t_2) + \Sigma (t_{Go} + t_{Gs}) + \Sigma t_{wt} [s]$$

167

where:

t_v = travel time
s = distance covered or angle of rotation per program step
v_{max} = maximum speed
t_1 = acceleration time
t_2 = retardation time
t_{Go} = time taken to open the gripper
t_{Gs} = time taken to close the gripper
t_{wt} = technologically conditioned waiting times for the robot (e.g. due to the opening of the chuck of a lathe)

The acceleration and retardation times t_1, t_2, and the maximum travel speed v_{max} are equipment-dependent values. The following approximate values may be assumed for industrial robots based on investigations into robots on the market (position in July 1978):

Maximum speed of translation:
$v_{max} = 1000$ mm/s

Maximum speed of rotation:
$v_{max} = 90°/s$

Acceleration and retardation time:
$t_1 + t_2 = 0.5$ s

When estimating the times required to open and close a mechanical gripper, the following assumption can also be used:

$t_{Go} \approx t_{Gs} \approx 0.3$ s.

Generally speaking, an industrial robot is used at a workplace to operate one machine or for the loose or rigid linking of seveal machines. In this case the production intensity of the workplace may be identified by the total quantity output M_F. M_F can be calculated by means of the calculated travel time, according to the following equation:

$$M_F = \frac{1}{t_v + t_w} \left[\frac{1}{s} \right]$$

In this equation t_w is the process-dependent waiting time for the robot. In the case of one-machine operation and a rigid connection of several production units, t_w is calculated as the sum of the main, secondary and **actual** distributing times. With a loose connection the calculation of t_w is more complicated, and is carried out by means of **queue theory** (5.14), or by simulating the operating problem on a digital computer system (5.15).

Much rarer is the case where an industrial robot operates several machines which work independently of each other. The main obstacle such an application lies in the lack of flexibility of the gripper systems currently available. For such a workplace, one value is insufficient to identify the output. In this case it is necessary to calculate the output for each production unit, one precondition for this being the division of the total travel time into operating times for each workplace element.

If the calculation of the quantity output(s) in relation to a planned layout shows that the output rate of the automated workplace does not reach the level of the manually operated workplace for any of the gripper designs tested, the implementation of the planned solution must generally be shelved. Only when the difference between the outputs of the automated and manual workplaces is very

168

small, or when a drop in production is expected, can a solution such as this be considered. Otherwise it must be determined whether the use of a second industrial robot or the dispensing of automation for individual production units represent suitable technical and economic solutions.

5.4.2.5 Selection of suitable industrial robots and peripheral equipment

5.4.2.5.1 Survey

The required characteristic features of suitable industrial robots and peripheral equipment can be analysed on the basis of the machine layout selected and the data obtained in the workplace analysis for a particular workplace. These requirements are combined with proposals regarding equipment designs in performance specifications. Suitable equipment can then be selected from those obtainable on the market on the basis of these specifications. In this case the market surveys on industrial robots and feeding devices contained in (5.16) and (5.8) may prove a valuable aid.

In addition to equipment selection, the performance specifications can also be used as a basis for the construction of equipment not offered on the market.

Furthermore, they form an important aspect of obtaining quotations for individual units selected for a short list.

5.4.2.5.2 Preparation of performance specifications for industrial robots

The scope of the performance specification for industrial robots depends essentially on the set of tasks to be performed. In the extreme case the performance specification may cover 80 different requirements, because as a study has shown (5.4), this number of features is necessary to describe all the characteristics of an industrial robot. Generally, however, the performance specification contains much fewer requirements. A selection of the most important requirements is given in Figure 5.32 and will be explained in the following:

– Usable workspace:
  The shape of the workspace of an industrial robot is determined by the number, type and arrangement of its basic axes. Its size and the points to be approached during a cycle are identified by an indication of the required travel distances, ranges and positions per axis.
– Gripper axes:
  Gripper axes are required to rotate the objects to be handled into the required positions, and to perform small translational movements. In addition they are necessary if, in order to shorten the non-productive times, two or more grippers are used connected by axes which are not functionally necessary.
– Load capacity:
  The load capacity is calculated from the maximum possible workpiece weight plus the estimated gripper weight. If several parts to be handled must be moved simultaneously, the sum of the individual weights gives the workpiece weight to be considered.
– Speed:
  Specific requirements with regard to speeds should only be imposed when functionally conditioned maximum speeds occur in the workplace studied. This is the exception, however. Generally speaking, the only requirement with regard to a workplace is that the movements be performed as quickly as possible. In this case, the robot selection should initially be made independently of the travel speeds. The attainable travel speeds should then be calculated on the basis of the specific equipment data for the solutions chosen. The outputs that can be

169

achieved can then be determined on the basis of these results, and a more detailed equipment selection made by comparison with the actual output of the existing set-up.
– Practicable scope of functions
The scope of the functions is defined by four items relating to equipment control:
1. nature of the control system required
2. program length
3. subprograms
4. linking capability

The types of control systems are divided into point-to-point and continuous path (PTP, CP) control systems. The functional difference between these two has already been explained in Section 2.3. The decision as to which is required for a particular task can be based on an analysis of the existing set-up.

The program length for continuous path control systems is given in minutes and for point-to-point control systems in the number of program steps. Whilst the program length of a continuous path control system is determined from the cycle time of the handling process, it is determined for point-to-point control systems from the number of positions to be approached during the operating cycle.

For sections which form part of the travel of an industrial robot
– which are repeated several times during a cycle,
– which are independent of preceding or succeeding process steps (example: loading and unloading of an unconnected machine in multi-machine operation)
– which are not continuously executed (example: spraying of dies after every 5th machining process)
– which experience regular changes (example: palleting of workpieces)

it is often appropriate, because of the reduction in storage requirement and programming effort as a result of this, to have subprograms available which can be called up at any time. The suitability of these subprograms may be determined from the analyses of the cycles of travel.

Industrial robots always operate together with one or more production units. For co-ordination of the times of the processes and in order to monitor certain conditions, the control systems of all the elements of the workplace concerned must be connected, i.e. interlinked. A sufficient number of signal inputs and outputs must be provided for this purpose in the control system of the handling equipment. Their number is determined from the functional cycle and the number of control devices provided.
– Other characteristics
All the other characteristics contained in Figure 5.32 follow directly from the workplace analyses. The type of gripper required for a particular workplace is determined by a comparison of the gripper designs and characteristics with the workpiece properties according to the criteria indicated in Section 2.4. Indications of the protective measures required against environmental influences are obtained from an evaluation of the working conditions, the installation position of the industrial robot being determined when establishing the machine installation plan.

5.4.2.5.3 Preparation of performance specifications for feeding devices
The characteristics listed in Figure 5.33 will generally suffice for the purpose of selecting the most important feeding devices now being offered on the market (5.8).
Many of these characteristics can be determined on the basis of the analysed

CHARACTERISTIC	EXAMPLE
Usable workspace	
– axes of rotation	C
– axes of travel	X, Z
– swivel angle (A, B, C) (°)	– ,–, 240
– travel distances (X, Y, Z) (mm)	1200, –, 800
– range (X, Y, Z) (mm)	2000, –, 1600
– approachable positions	
(X, Y, Z, A, B, C)	40, –, 5, –, –, 40
Gripper axes	
– rotational axes	A, C
– translational axes	Y
Load capacity (kg)	50
Positioning error (mm)	1, 0
Speed	
– of the translational axes, X, Y, Z	
(mm/s)	1000, –, 500
– of the rotational axes A, B, C (°)	–, –, 90
Practicable scope of functions	
– type of control system	point-to-point
	control system
– program length	60 program steps
– number of subprograms	3
– number of required signal	
inputs and outputs	15, 10
– gripper	double gripper
Protection against environmental	encapsulation as
	protection
influences	against heat
Installation position	overhead

Figure 5.32   Important characteristics for selecting suitable industrial robots.

workpiece spectrum or the system solutions developed. Others can be derived from the geometrical conditions. Thus the dimensions and direction of the feed channel is determined essentially by the position of the input and output channels of the machines and by the chosen machine arrangement.

When preparing the performance specifications, problems only generally arise in specifying the output data. Thus the feed rate must be chosen so that it guarantees the attainment of the maximum possible output in the automated condition. For this purpose it is necessary to calculate the individual outputs that can be obtained referred to the industrial robot types selected. Apart from the output, the minimum feed rate required also depends on the maximum workpiece length in the direction of feeding. Furthermore, the distances required between the workpieces during the feeding process must be taken into consideration when determining the feed rate in cases where the parts cannot be fed without being separated by a gap. Finally, the feed rate in relation to the ordering devices is still determined essentially from the probability of being in order, i.e. the ratio of the workpiece position favourable for the sorting process to the total number of possible workpiece positions (5.17, 5.18).

171

All the other characteristics listed in Figure 5.33 may be determined essentially on the basis of the analysed environmental conditions. A more detailed specification of the drive is only required when not all drive types can be used because of certain environmental conditions (e.g. explosion risk). In designing the accessories, particular attention must be paid to the required protective measures against unfavourable environmental influences (dirt, dust, high temperatures).

Furthermore, particular attention must be paid to additional equipment which is required for monitoring fault-free operation (hold-up protection, filling level monitoring).

5.4.3 Establishment of the optimum overall solution

Several technical implementation possibilities may be found for the individual elements of the system solution developed on the basis of performance specifications for industrial robots and feeding devices. It is necessary to select the optimum partial solution from these possibilities in the next planning stage. For this the conducting of an efficiency analysis is again recommended, generally using the same criteria selected for evaluating the system solutions. By combining the optimum partial solutions the optimum technical overall solution is finally obtained for automating the given workplace.

5.4.4 Considerations of cost effectiveness

The decision whether to implement the solution arrived at in the above planning stage or not can only be taken on the basis of a cost effectiveness study. In this study, the costs of the manual actual condition must be compared with the costs of the automated workplace (assumed condition). Furthermore, a variation in

CHARACTERISTIC	EXAMPLE
Principle of function – Characteristic feed function – Characteristic function principle	Storage, transfer, orientating Vibratory bowl feeder
Workpiece data – Workpiece group – Workpiece weight (kg) – Workpiece dimensions (mm)	Block part 0.075 $100 \times 20 \times 10$
Output data – Feed rate (m/min) – Storage capacity (1) – Load capacity (kg)	12.0 7.0 15.0
Feed channel – Direction – Width (mm)	horizontal 20
Equipment dimensions (diameter × height) (mm)	300×320
Type of drive	electromagnet
Accessories required	accumulation protection

Figure 5.33   Important criteria for selecting feeding devices.

172

output must also be considered if the workplace concerned can be fully loaded in the event of increased output.

The following four charcteristics have proved suitable in practice for assessing the feasibility of an investment:
- cost saving
- amortization
- profitability
- limiting production quantity

Static and dynamic computing methods can be used to calculate these characteristic values. The static methods use average values, i.e. the characteristic cost effectiveness values are only calculated for a certain period. On the other hand, the dynamic methods take into consideration the difference in time in the occurrence of expenses and receipts connected with an investment project, i.e. the interests relating to expenditure and receipts during the effective investment period are taken into consideration.

The dynamic methods impose stringent requirements on data collection and only afford advantages when the costs incurred and the receipts obtained can be predicted for the whole effective period to a high degree of accuracy.

As this is not yet possible for an industrial robot application due to lack of experience, it does not seem appropriate to use dynamic computing methods for these tasks. In the following, therefore, only the static methods will be given further consideration.

5.4.4.1 Survey of the methods of cost effectiveness calculation

5.4.4.1.1 Calculation of cost saving

The cost saving is determined on the basis of a cost comparison calculation. If two projects with the same capacity (e.g. pieces/month) are compared, a simplified space cost calculation can be carried out. If the systems to be compared have different capacities, a piece cost comparison (5.19) must be carried out and the limit piece number (see Section 5.4.4.1.4) determined.

The following types of cost are of importance for the comparison calculation:
Fixed cost:
- Calculated depreciation:

$$\text{Calculated depreciation} = \frac{\text{original value}}{\text{technical effective period}}$$

When estimating the calculated depreciation, a general increase in prices of capital plant can be considered, where it is not the original value of the plant but its replacement value that is used as the initial basis for the depreciation rates.

- Calculated interest:

$$\text{Calc. interest } (Z) = \text{Calc. interest rate} \times \text{average capital investment}$$

The calculated interest rate is equivalent to the interest rate for long-term capital.

In the case of linear depreciation, the average capital investment is 50% of the replacement value.

- Floorspace costs

Variable costs:
- Wage costs

These are composed of the
- production wage costs
- setting up wage costs
- general personnel costs

These are composed of the
- managerial costs
- general material costs
- energy costs
- maintenance costs
- other variable costs

The cost saving (K_e) is obtained from the difference between the sums of the fixed and variable costs of the projects to be compared.

$$\text{Cost saving } (K_e) = (\Sigma \text{ fixed costs} + \Sigma \text{ variable costs})_I \\ - (\Sigma \text{ fixed costs} + \Sigma \text{ variable costs})_{II}$$

5.4.4.1.2 Calculation of the amortization period

The amortization calculation (also termed capital return calculation) determines the period required for the invested capital to be recouped by the Company through income. The amortization period is calculated from:

$$\text{Amortization period} = \frac{\text{Capital invested}}{\text{Average capital return}}$$

The average capital return is obtained from the profit (corresponding to the annual cost saving (K_e)) and the calculated depreciation.

The following is then obtained:

$$\text{Amortization period} = \frac{\text{Capital invested}}{\text{Annual cost saving} + \text{depreciation}}$$

The amortization period is an essential parameter for assessing the risk incurred by the investment. The shorter the amortization period the smaller the risk. From this follows the pre-condition for a correct investment:
Amortization period < useful life.

5.4.4.1.3 Calculation of profitability

Profitability is generally defined as the ratio of "profit from investment" to "capital invested".

$$\text{Profitability} = \frac{\text{annual cost saving}}{\text{average investment of capital}} \times 100$$

The cost saving is in this case the saving resulting from the replacement of the old

174

system by a more rational system. The capital invested is the additional capital required for realising the object of the investment being considered.

5.4.4.1.4 Calculation of the critical production quantity
The critical production quantity may be determined by means of the limiting quantity calculation 5.20. This is necessary when determining the more cost-favourable alternatives by a unit cost comparison for variable production capacity.

The determination of the critical production quantity is particularly necessary when the planned output is difficult to estimate. With the limiting quantity calculation, the output at which a change should be made from one process to another can then be established so that production can always be carried out under the most favourable cost conditions.

5.4.4.2 Workplace costs incurred in the industrial robot application
To ensure that all the data relevant to the investment are collected, the use of checklists is recommended. Such a checklist for determining all the cost types to be taken into consideration in an industrial robot application is shown in Figure 5.34 (5.21). The following explanatory notes are considered necessary for the following lines of the form:

(1) The purchasing costs include the costs of the basic unit without special grippers and the expenditure incurred for freight, insurance and packing.
(5) The planning costs consist of costs for in-house designs and for outside plans and expert reports.
(7) The calculated interest rate may be estimated at 10% if no better reference values are available.
(9) No comprehensive experience has so far been gained of the useful life of industrial robots. Many users tend to want to pay off this equipment in one to a maximum of two years. This attitude, which derives from the use of special machines and rigid automation devices, is certainly wrong for flexible industrial robots. An estimate of a useful life of at least three years, or in most cases even five years, is justified all the more, the more flexible the equipment concerned is and the more loosely it is linked to the workplace or the operation in question.
(12) The setting costs are calculated as the product of the setting frequency, average setting up time and the wage costs for tooling up. The resetting time is in this case obtained from the number of different production batches per annum.
(13) The programming costs are, like the setting costs, calculated from the frequency of the programming process, the average programming times and the wage costs for the programmer. The number of new programming operations does not always coincide with the number of resetting processes. Thus in the case of equipment which offers the possibility of storing programs that have been established on data carriers, new programming is only necessary when new products appear. The precondition in this case, however, is that the position of the tools can be reproduced exactly after setting.
(14) The production wages are not omitted altogether in most cases and a certain proportion must continue to be taken into account for monitoring tasks and auxiliary functions (e.g. fitting of magazines).
(15) The energy costs are calculated from the installed power, the time it is switched on and the load factor. Suitable estimates are required for this, but experience shows that it is not a determining cost factor.
(16) Work area costs need only to be taken into consideration when the space

175

Item	Data			Calculation
	Designation	Abbreviation	Unit	
(1)	Procurement costs, basic unit	K_{G1}	DM	
(2)	Costs of periphery (grippers, sensors, feeding devices)	K_{G2}	DM	$K_{GES} = \sum\limits_{i=1}^{5} K_{Gi}$
(3)	Machine modification costs	K_{G3}	DM	
(4)	Installation costs	K_{G4}	DM	
(5)	Planning costs	K_{G5}	DM	$Z = \dfrac{K_{GES}}{2} \cdot iZ$
▶ (6)	Capital investment	K_{GES}	DM	
(7)	Calculated interest rate	iZ	%/a	$A = \dfrac{K_{GES}}{t_N}$
(8)	Calculated interest	Z	DM/a	
(9)	Economic useful life	t_N	years	
(10)	Calculated depreciation	A	DM/a	$K_{Fix} = A + Z$
▶ (11)	Sum of fixed costs	K_{FIX}	DM/a	
(12)	Resetting costs	K_{UM}	DM/a	
(13)	Programming costs	K_{PRO}	DM/a	
(14)	Production wage costs	K_{FE}	DM/a	
(15)	Energy costs	K_{EN}	DM/a	
(16)	Floor space costs	K_R	DM/a	$K_{Var} = K_{UM} + K_{PRO} + K_{FE}$
(17)	Maintenance costs	K_{IN}	DM/a	$+ K_{EN} + K_R \quad + K_{IN}$
▶ (18)	Sum of variable costs	K_{VAR}	DM/a	$K_{IR} = K_{FIX} + K_{VAR}$
▶ (19)	Annual costs for IR.	K_{IR}	DM/a	

Figure 5.34   Checklist for determining all the costs incurred in the industrial robot application.

required for the automated workplace differs from that of the manual workplace owing to machine relocations.

(17) Not enough experience has yet been gained of the maintenance costs of an industrial robot. To estimate these costs values gained from experience for production equipment should be used as a basis. These are between 5% and 10% of the procurement costs.

5.4.4.3   Procedure for cost effectiveness calculation

A proposal is shown in Figure 5.35 for making the cost effectiveness calculation for an industrial robot investment.

First all the costs which are incurred at the given workplace in the manual and automated condition must be established and quantified. As far as the automated condition is concerned, these costs may be determined on the basis of the checklist given in Figure 5.34. The following costs must be determined for the manually operated workplace:

– operating wage costs
– maintenance wage costs
– instruction wage costs
– costs of product changeover
– work area costs

After recording the data, the cost saving, profitability and authorisation period should be calculated in succession. If no clear decision for or against the industrial robot application is possible on the basis of these results, an attempt must then be made to estimate the investment risk more accurately by means of a sensitivity analysis.

176

Figure 5.35
Procedure for carrying out
the cost effectiveness calculation
for an industrial robot
investment.

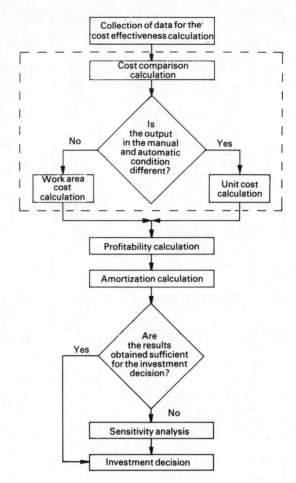

This sensitivity analysis enables the influence of uncertain factors on the feasibility characteristics to be determined.

It can provide the answer to the following questions:

(a) How far can one or more input factors deviate from their original values without the result dropping below or exceeding a fixed value?

(b) How does the result of the calculation vary when one or more input values are varied by a certain amount.

By contrast to other methods with a similar objective (risk analysis, methods for forming the weighted mean value), the sensitivity analysis is very simple to carry out. If the first question has to be answered, the procedure consists of the following steps:

(1) Selection of the values considered to be uncertain, e.g.:
  – maintenance costs
  – machine modification costs
  – setting costs

(2) Determination of the functional dependence between the output value (e.g. amortization period) and the selected input value.

177

	DESCRIPTION	UNIT	NUMERICAL VALUE		
			Original estimate	Variation I optimistic	Variation II pessimistic
COSTS IN IR APPLICATION	Original costs (industrial robot and periphery)	DM	260,000	260,000	260,000
	Machine modification cost	DM	120,000	80,000	160,000
	Installation costs	DM	20,000	10,000	30,000
	Planning costs	DM	40,000	40,000	40,000
	Original capital investment	DM	440,000	390,000	490,000
	Average capital investment	DM	220,000	195,000	245,000
	Calculated interest rate	%/a	10	10	10
	Calculated interest	DM/a	22,000	19,500	24,500
	Economic life	years	5	5	5
	Calculated depreciation	DM/a	88,000	78,000	98,000
	Sum of fixed costs	DM/a	110,000	98,000	122,000
	Resetting and programming costs	DM/a	2,000	2,000	2,000
	Production wage costs	DM/a	75,000	75,000	75,000
	Energy costs	DM/a	1,500	1,500	1,500
	Additional work area costs	DM/a	1,000	1,000	1,000
	Maintenance costs	DM/a	25,000	13,000	39,000
	Sum of the variable costs	DM/a	105,500	92,500	118,500
	Annual costs of IR	DM/a	215,500	190,500	241,000
COSTS FOR MANUAL OPERATION	Operating wage costs	DM/a	300,000	300,000	300,000
	Maintenance wage costs	DM/a	6,000	6,000	6,000
	Instruction wage costs	DM/a	1,000	1,000	1,000
	Costs of product changeover	DM/a	1,000	1,000	1,000
	Annual costs for manual operation	DM/a	308,000	308,000	308,000
YIELD RATIO	Output with manual operation	Stck/a	60,000	60,000	60,000
	Output with IR application	Stck/a	60,000	60,000	60,000
	Output ratio (manual/IR application)	–	1	1	1
COST EFFECTIVENESS DATA	Cost saving (when changing over to IR use)	DM/a	92,500	117,500	67,000
	Profitability	%/a	42.0	60.2	27.3
	Amortization period	years	2.4	2.0	3.0

DM/a = Deutschmarks per annum
Stck/a = pieces per annum
*varied values

Figure 5.36   Example of a cost effectiveness calculation for an industrial robot application (Part I).

(3) Breakdown of the function according to the input value studied for the established limit of the output value.

To answer the second question all that is required is to carry out the original investment calculation again once or several times, varying the data considered uncertain. The data can in this case be varied by fixed percentages, or upper and lower limit values can be used for the data selecting optimistic or pessimistic estimates. The example below (Figures 5.36 and 5.37) illustrates the function of the sensitivity analysis and the calculation processes to be carried out before the analysis.

178

Question: What are the maximum machine modification costs (K_{G3}) if the amortization period is 3 years maximum?

Solution: Solve the equation for the amortization period according to the machine modification costs.

Amortization period (years): $3 = \dfrac{\text{Capital investment}}{\text{calculated depreciation} + \text{cost saving}}$

Capital investment (DM): $K_{GES} = 320{,}000 + K_{G3}$

Calculated depreciation (DM/a): $A = 64{,}000 + \dfrac{K_{G3}}{5}$

Cost saving (DM/a): $K_{SPAR} = 308{,}000 - (64{,}000 + \dfrac{K_{G3}}{5} + 16{,}000 + \dfrac{K_{G3}}{20} + 105{,}000)$

$$\Rightarrow 3 = \frac{320{,}000 + K_{G3}}{64{,}000 + \dfrac{K_{G3}}{5} + 122{,}500 - \dfrac{K_{G3}}{4}}$$

$$\Rightarrow 3 = \frac{320{,}000 + K_{G3}}{186{,}500 - \dfrac{1}{20} K_{G3}}$$

$$\frac{23}{20} K_{G3} = 559{,}500 - 320{,}000$$

$$K_{G3} = 208{,}261$$

The machine modification costs must therefore not exceed DM 208,261.

Figure 5.37   Example of a cost effectiveness calculation for an industrial robot application (Part II).

5.5   Implementation of the chosen concept

If the cost effectiveness of the industrial robot application has been demonstrated on the basis of the previous investment calculation, it is then necessary to plan the implementation of the chosen concept. For this purpose all the individual tasks to be performed must first be analysed and their chronological and logical sequence determined. As a result of this investigation a flowchart can be drawn up to serve as a guide to further procedures. Such a network plan is shown in Figure 5.38, with the most important tasks to be performed in implementing the solution concept. It will be clear that until the industrial robot is finally incorporated in the manufacturing process an abundance of problems, some of which are very complex, remain to be solved.

The first step is to subject the characteristic data for the equipment selected to a critical comparison with the workplace requirements once again, in order to establish any equipment modifications or special designs that may be required. The orders for the equipment concerned must then be written out on the basis of this investigation. In parallel with this, it must be determined whether the alterations and adaptation work required can be carried out at the workplace by departments in the company itself, or whether it is advisable to have this work carried out by outside firms. It must also be ascertained whether the modifications required should be passed on to outside firms. To answer these questions the capacities of one's own firm must be assessed and outside quotations obtained. Finally, as a

179

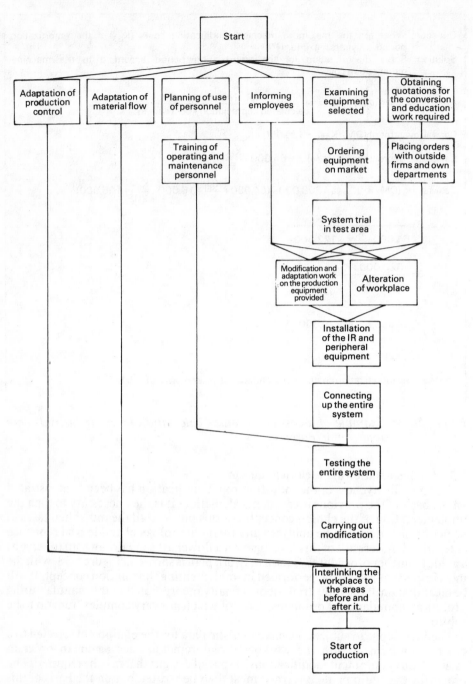

Figure 5.38: Flowchart plan for implementing an industrial robot application.

result of these investigations, the optimum departments for the work in question can be chosen and given the task of carrying out the work.

Delivery of the industrial robot and peripheral equipment may be followed by a test on the equipment on the test stand, which should be carried out in two stages.

First it should be determined, as part of an acceptance test, whether the characteristic values promised by the manufacturer are actually attained. To do this the robot is tested for the positioning accuracy and attainable speeds. Suitable procedures for these tests are described in Chapter 3. The robot should then be connected to the peripheral equipment to test the behaviour of the entire system. In doing so, the connection to the production units must be simulated by suitable additional equipment or testing devices.

Testing on the test stand is advisable when the application is a new one, from the technical point of view, where long installation and test times can be expected. This testing may be dispensed with if the loss of production caused by installation of the handling equipment can be successfully compensated for by pre-production at the workplace to be automated or by supplementary production capacity (e.g. change from 1-shift to 2-shift operation). A production loss may also be prevented if works holidays are used for the installation.

After testing on the test stand, which may also lead to modifications to the designed layout, work can be begun on the conversion. This work may extend in particular to the following areas:

– installation of the energy supply system
– conversion of the vacuum system
– preparation of new machine foundations
– reinforcement of floor and ceiling supports

In parallel with this conversion work, the required modifications can be carried out on the production equipment (installation of control devices for monitoring the operating process, subsequent automation of the auxiliary material supply, waste removal, etc.).

In the subsequent installation of the industrial robot and the peripheral equipment, the linking of the subsystems and their connection to the production equipment and feeding devices present the main problems.

The whole automated system must then be tested under production conditions. All breakdowns and failures must be recorded. The error sources must then be analysed on the basis of the reports and suitable modification measures worked out.

In addition to the technical tasks listed, important organisational tasks must also be carried out at the workplace operated by the industrial robot until production is started. Thus suitable measures must be taken in production control and material flow planning where an increase in output can be achieved by automation. Similarly, modifications must be carried out in these areas if the workpiece range machined at the automated workplace no longer corresponds to that in the manually operated workplace.

In addition to these organisational modifications, the use of personnel at the automated workplace must be planned and suitable training of operating and maintenance personnel undertaken. Finally it is also necessary to dispel any objections the employees may have to the use of the industrial robot in informative discussions.

As can be seen from the above, the implementation of an industrial robot application is a highly complex task. For this comprehensive specialist knowledge is required and only in exceptional cases is this available in the user firm. The implementation must therefore be undertaken jointly, in most cases, by the user and the equipment manufacturer. In addition, it is often advisable to employ consulting engineers or research institutes for solving special problems.

5.6 Application examples

5.6.1 Survey of industrial robots used

Figure 5.39 gives a survey of industrial robots in use in the German Federal Republic at the end of 1977, classified according to areas of application. The percentage of different applications will vary only slightly in the coming years.

5.6.2 Industrial robots for tool handling

5.6.2.1 Industrial robots for coating

Together with spot welding, coating (e.g. application of paint, enamel, underbody protection) is one of the classic applications of industrial robots. The robot has the task of performing defined movements in the working area with a spray gun.

The equipment often designed for this purpose is in most cases programmed by the "teach in" method (see Section 2.3.3.2.1 Programming Processes), i.e. the movements are carried out by a "programmer" and are reproduced by the robot. Special programming knowledge is not required of the operator. One disadvantage of equipment now on the market in particular is that the operator must move not only the paint spray gun but also the arm of the robot necessitating possibly several attempts to achieve an adequate painting quality. The equipment is expected to reproduce the programmed path as accurately as possible at the same speed. Compared with other tasks, however, the required positioning accuracy is very low.

The handling weight is low (spray-gun) but varying forces also occur caused by paint supply hoses and programming forces in the teach-in operation. Since the parts to be coated are mostly in motion (conveyor belt, overhead conveyors), the robot must be capable of following this movement continuously.

Greater synchronous movement accuracy can be achieved by means of an external measuring system on the conveyor.

	Application	Number of units	% proportion
Tool handling	Coating	90	17.3%
	Spot Welding	74	14.2%
	Seam welding	12	2.3%
	Trimming	10	2%
	Assembly	2	1%
	Other	2	1%
	Total	ca. 190	36.5%
Workpiece handling	Handling on presses	50 (70)	9.6%
	Handling on forging machine	30 (30)	5.8%
	Handling pressure on casting machine	60 (190)	11.5%
	Handling on die casting machine	60 (90)	11.5%
	Handling generally	100 (120)	19.2%
	Total	300 (500)	57.7%
Research purposes		16	3.1%
Applications not known		15	2.9%

Figure 5.39   Number of industrial robots in use according to applications.

182

Task	Example No.
Glazing of washbasins	1

Statement of problems:

● Workpieces difficult to handle must be glazed from different sides.

● The glazing must be applied uniformly.

● It must be possible to glaze workpieces of different shapes and sizes over the entire installation without a major setting effort.

Solution:

Layout of the complete installation as shown in Figures 5.40 and 5.41.

● Programming is carried out at the beginning of the application, by the operator spraying the workpiece by means of the spray gun fitted onto the arm of the industrial robot. When this motion cycle has been completed it is stored and can then be repeated continuously by the robot.

● The workpieces are manually removed at the input and output stations and at these stations (untreated part) and (finished part), transferred to the robot by means of a circular roller conveyor system.

● After an untreated workpiece has been loaded, a starting pulse is automatically initiated by means of a switch. The robot then begins its work.

● The workpiece, which is placed on a revolving table, is then glazed automatically. The electrical signals for rotating the table are initiated by the industrial robot control system.

Task	Example No.
Glazing of washbasins	1

Auxiliary equipment:

To minimise the loss of glazing material the part which is sprayed past the workpiece is for the most part recovered by means of dry pre-separation on the spraying booth. Here the material sprayed against it runs off into a collection tank. The remainder is washed out with water and is deposited in the easily accessible rear section of the water tank.

Reasons for application:

● Elimination of inhuman workplaces

● Uniform output at max. speed

● Uniform application of the glazing

● Short setting time by "storing" the programs on magnetic tapes

Industrial robot used	Devilbiss-Trallfa
Cycle time (s)	30 to 60 (depending on workpiece size)
Weight on arm of robot (kg)	
Programming method	Teach-in
Shifts/day	2 to 3

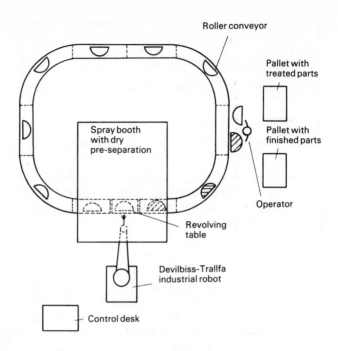

Figure 5.40
Complete layout of
the washbasin and
glazing plant.

Roller conveyor

Pallet with
treated parts

Spray booth
with dry
pre-separation

Pallet with
finished parts

Operator

Revolving
table

Devilbiss-Trallfa
industrial robot

Control desk

Figure 5.41   Roller conveyor system with Devillbiss-Trallfa industrial robot for
glazing washbasisns (Photo: Devillbiss-Trallfa).

185

Task	Example No.
Painting of engine blocks	2

Statement of problems:

● Movement of the complicated engine shape at a distance of approx. 150 mm, at a continuous belt speed of 5 m/min.

● Sufficient safety precautions taken to ensure that the operator is able to carry out the preparatory work such as covering the openings and finish-painting the engines whilst the robot paints one engine per pass.

Solution:

The arrangement of a robot for painting pre-positioned engines on an assembly line is shown in Figure 5.42. The programming is carried out as already described in Example 1. By contrast, however, storage is in a magnetic disc memory, so that the mumber of programs is limited to a maximum of 75, for a total running time of 900 s. The stored program can then be called up by pressing a button or coding, depending on the workpiece.

Operating cycle:

● Covering the openings on the pre-positioned engine by an operator.

● Painting about 60 to 70% of the engine surface with a coat 7 μm thick by the robot at a stady belt speed.

● Visual quality control and finish-painting by another operator.

Task	Example No.
Painting of engine blocks	2

Additional equipment

- The robot is fitted with an automatic electrostatic gun with air atomisation.

- The spray booth provided had to be extended so that the robot could be installed. The control system, with magnetic disc memory, and the hydraulic unit are located outside the booth.

Reasons for application:

- Elimination of inhuman workplaces

- Uniform quality

- Spray booths can be entered whilst the robot is in operation. This makes it possible to test and possibly correct the spray pattern.

Industrial robot used	SK-Coat-A-Matic (Retab)
Cycle time (s)	approx. 20
Weight on the arm of the industrial robot (kg)	up to approx. 15
Shifts/day	2 to 3
Number of programs	75

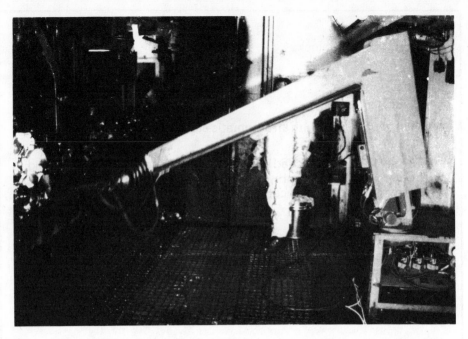

Figure 5.42   Painting of engine blocks (Photo: Spritztechnik Kopperschmidt).

5.6.2.2   Industrial robots for spot welding

Manual spot welding is one of the most strenuous physical tasks as the welder must be able to position a heavy spot welding elctrode accurately and at high speed. Because of the heavy weight to be handled (up to approx. 60 kg), the speeds required and the high positioning accuracies, the requirements imposed on these industrial robots are very stringent. As the welded spots are generally so close together that the unit does not reach its maximum speed, considerable accelerations and retardations are required to achieve the low predetermined cycle times. Robots for spot welding are characterised by a high degree of rigidity, strong drive systems, large work spaces and high load capacities. The possibility of synchronous processing with moving work carriers is partly available. Problems also arise with the welding current supply, as variable forces are frequently exerted on the robot due to rigid cables, in addition to the heavy weight of the welding electrode. The successful application of industrial robots in spot welding is due, as for coating, to the fact that no major costs are incurred for additional peripheral equipment, by contrast to manual or conventional automatic systems. In future, the robot will gain increasing importance in this field as an alternative to the conventional automatic welding line.

5.6.2.3   Industrial robots for arc welding

In arc welding, the industrial robot must guide a welding torch as accurately as possible along a programmed track. The programming may be carried out point by point, followed by interpolation between the points, or, as with coating, may be carried out using the teach-in method (see Chapter 2.3.3.2.1, Programming processes). Since the tolerances of the parts to be welded together are frequently very high, sensor-guided seam welding is gaining in importance. If the workpieces are very large, or are inaccessible for one welding cycle, a co-ordinated movement is

Task	Example No.
Spot welding of body parts	3

Statement of problems:

- Spot welding – cut-out in rear window.
- Spot welding – boot bottom joining.
- Spot welding – end piece/side section joining.
- Integration into existing gantry frame structure of the body welding line.
- Allowance for transport movement by means of longitudinal rail conveyor.

Solutions:

Figure 5.43 shows the diagrammatic arrangement of three robots used, in relation to each other and to the body. To ensure that the required facility of movement could also be guaranteed, two of the robots were installed suspended overhead, and one standing upright.

Operating cycle:

- The body conveyed by means of the longitudinal rail conveyor is raised to working height at the welding station, and positioned.
- Execution of the welding operations. All three industrial robots operate in parallel so that the three operations are carried out simultaneously.
- After the welding operation has been completed the robots return to their initial positions; the body is lowered and conveyed to the next station.

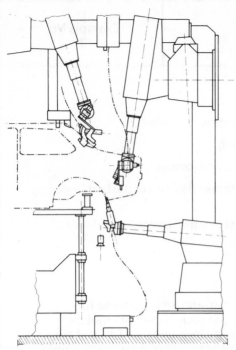

Figure 5.43
Layout of the three robots in a body
shell welding line station (Photo:
VW AG).

Figure 5.45
General view of the welding station
with three robots (Photo: VW AG).

Figure 5.44   View from the maintenance platform at the side during the welding
process (Photo: VW AG).

Task Spot welding of body parts	Example No. 3

Additional equipment:

- Development of spot welding electrode with which the welding process can be carried out on both sides of the body using a single unit.

- Protective enclosure of the entire system.

- Fixed working platforms for maintenance and tool changing.

Reasons for application:

- Cost saving over fully mechanised systems.

  Fully mechanised spot welding plants are required for the series production of large quantities. However, the expenditure relating to these plants is very cost-intensive, and in addition the plant must be set for every part modification, no matter how small.

- The displacement of spot welds or the addition of welds can be achieved without difficulty by program modification.

Industrial robot used	Röhren-Gerät R 30 (VW AG)
Cycle time (s)	26
Drive	Electromechanical
Shifts/day	2
Number of welded spots	48
Type of control system	Point-to-point
Programming method	Teach-in method
Number of axes	3 primary axes, 3 manual axes

required between the robot and the workpiece (or workpiece fixture). The synchronous travel accuracy achieved determines the quality of the welded seam. Industrial robots which are particularly suited for track welding are characterised by a high degree of path accuracy and elaborate kinematics (5 to 6 axes). The handling weight is determined by the weight of welding torch; additional forces may occur by welding gas and welding wire supply, or by welding current supply in the case of arc welding.

5.6.2.4 Industrial robots for fettling

The fettling of workpieces is one of the operations which more than any other imposes strain on the human operator, e.g. heavy weights, loud noise and dirt, wearing of protective clothing (breathing mask, goggles, hearing protection). The low incidence of industrial robot applications for this work is due to the fact that effective parameters of, for example the irregular flash geometry encountered

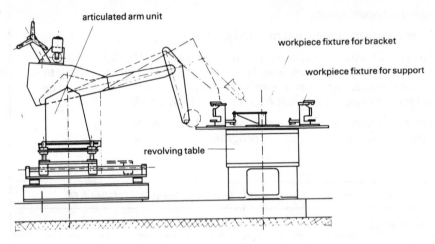

Figure 5.46   Arrangement of the entire inert gas-shielded welding plant (Photo: VW AG).

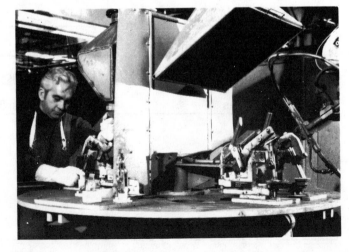

Figure 5.47
Revolving with four
working stations
for the inert
gas-shielded
welding of two
different workpieces.
(Photo: VW AG).

192

Task	Example No.
Welding of thin sheet metal	4

Statement of problems:

● Welding thin sheet metal with better form-holding and greater durability.

● System must be usable for different parts.

Solution:

Two different workpieces, bracket and support, can be welded on one revolving table at four stations, Figure 5.46.

However, only one assembly, of either the bracket of support, is carried out per shift. There setting of the table and robot is always carried out at the beginning of the shift.

Operating cycle:

● The operator places the component parts of the part to be welded on the fixture designed for this purpose.

● When all the parts are in position, the workpiece is fed to the robot by turning the table.

● Whilst the robot is carrying out the welding process, the operator removes the welded workpiece and places the next component parts on the table.

Additional equipment:

● Revolving table divided into four sections for receiving and feeding the workpiece fixtures to the robot.

● Protective fencing with a protective window and door.

Task Arc welding of thin sheet metal	Example No. 4

Reasons for the application:

● Humanisation of the workplace by isolating the welding area; as a result the operator is no longer exposed to the effects of heat and fumes during welding.

Industrial robot used	Articulated arm unit K 15 (VW AG)
Cycle time (s)	26.6
Shifts/day	2
Drive type	Electromechanical
Type of control system	Point-to-point and continuous path control system
Programming method	Teach-in method
Number of axes	3 primary (main) axes, 2 manual axes 1 travelling unit

Task	Example No.
Arc welding of punched sheet metal parts	5

Statement of problems:

● Welding of pressed and punched sheet metal parts using inert gas (car seat)

Solution:

Figure 5.48 shows a view of the whole system. The sheet metal parts are clamped by an operator in a welding frame at the feed stations. To prevent the operator from being injured during the welding process, the positioning takes place behind a protective screen. The use of a revolving table, on which two clamping devices are mounted, was therefore necessary. These are staggered by 180°. Both fixtures are rotated alternately from the loading to the welding station (Figure 5.49).

The welding process is carried out by the robot used in three planes so that the clamped workpiece position can be retained.

Additional equipment:

● Clamping fixture for the sheet metal parts to guarantee dimensional stability.

● Rotating device for turning the preclamped sheet metal parts to the welding area.

Task	Example No.
Seam arc welding of punched sheet metal parts	5

Reasons for the application:

- Uniform quality.
- Subsequent treatment no longer required, since the welded seam is clean and uniform.
- Elimination of an inhuman workplace, since despite screening off and removal by suction of the heat and fumes developed put a strain on the welder.

Industrial robot used	ASEA b 6
Cycle time (s)	90
Programming method	Teach-in method
Shifts/day	2
Welding speed (m/min)	5
Type of drive	electrical

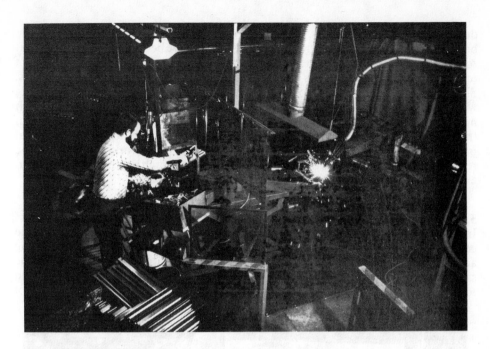

Figure 5.48   General view of the workplace (Photo: ASEA).

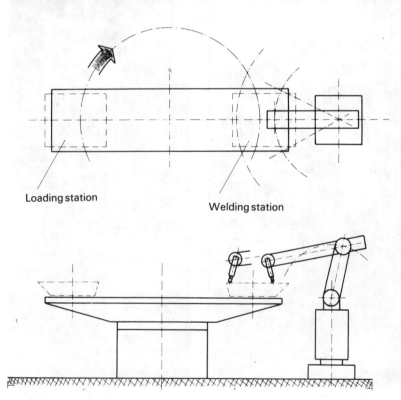

Loading station

Welding station

Figure 5.49   Layout of the whole system.

during the working process, cannot be accurately predetermined. Thus the breakthrough is not expected until suitable sensors have been developed. Industrial robots which are suitable for these tasks must be capable of calculating their path of movement from sensor signals and correcting it continuously during the process. Since the industrial robot has to guide driven tools (cutters, grinders), the handling weights are heavy. In addition, forces are produced by the machining process which necessitate considerable static and dynamic rigidity of the robots. Due to the size of the workpieces and the complex path to be travelled during deburring, essential preconditions for the use of industrial robots are a large working area and the requisite number of degrees of freedom (5 to 6 axes). A high degree of resistance to extreme environmental influences is a further precondition.

5.6.2.5 Industrial robots for assembly
Industrial robots are still used least for assembly, which is due to the considerable complexity of the operations. Maximum demands are made on positioning accuracy, particularly in the execution of mating operations, which can only be satisfied by providing a control system with sensors. Since different workpieces and tools have to be manipulated during an assembly process, a high degree of

Figure 5.50   General view of the workplace (Photo: ASEA).

198

Task	Example No.
Fettling of castings	6

Statement of problems:

● Removal of sprues, risers, etc. from castings after ejection from the casting mould.

● Removal of casting flash on the mould joint.

Solution:

The castings are picked up manually and clamped in fixtures. The workpieces to be fettled are fed to the industrial robot by means of a revolving table, the robot being installed in a separate booth away from the operator (Figure 5.50). Two fixtures displaced by 180° are mounted on the table so that the workpieces can be placed alternately on them whilst the robot performs the fettling process with a grinding or cutting-off wheel. When the workpieces are changed over, the fixtures on the table must also be changed. The corresponding program for the robot must also be read in.

Additional equipment:

● Protective booth – the robot was installed in a protective booth to prevent the operator from being troubled by sparks, grit and noise.

● Revolving table owing to the separation of the loading and processing stations these must be connected by means of a revolving table.

Task Fettling castings	Example No. 6

Reasons for the application:

● Removal of an inhuman workplace.

● As a result of the mechanical parting off of the sprues and risers some of the additional machining can be dispensed with.

Industrial robot used	ASEA
Cycle time (s)	65
Shifts/day	2 to 3
Grinding speed (m/s)	90 to 100

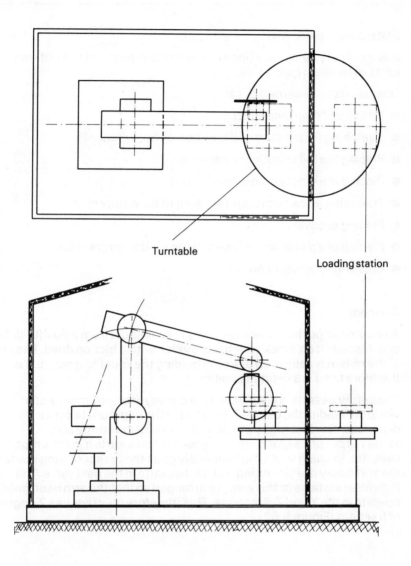

Turntable

Loading station

Figure 5.51   Layout of the whole system.

201

| Task | Example No. |
| Assembly of typewriter ribbon cassettes | 7 |

Statement of problems:

Assembly of typewriter ribbon cassettes composed of five different parts, in medium quantities.

Operations to be carried out are:

- Loading of empty cassette
- Searching for the start of the ribbon on the full spool
- Placing the full spool in the cassette
- Placing the empty core in the cassette
- Threading in the ribbon and sticking to the empty core
- Placing of cover
- Placing of a piece of cardboard for unwinding protection
- Pressing together of housing

Solution:

A pneumatic pusher device removes the full spool from a double slide and deposits it on a disk which is provided with a friction drive. The end of the ribbon is caught up at a stop holding the spool as a result. It is therefore fixed in a definite position.

The empty cassettes are placed on a conveyor belt by means of a vibratory feeder, the belt being loaded at the appropriate cycle distances. The first (right-hand) arm removes the empty core at the output of a second vibratory feeder and inserts it in the empty cassette. In doing so it simultaneously grips the full spool lying ready on the stationary disk, inserting it in the cassette in the next cycle. The individual stations of this arm are arranged so that the arm need only operate in the Y and Z directions. This therefore only requires 2 degrees of freedom (Figure 5.52).

Figure 5.52   *Working area of the 1st arm with magazine for empty core, spool and housing (Photo: Olivetti).*

The typewriter ribbon cassette is then fed to the working area of the third arm, where the cover positioned by a pneumatic transfer unit from a stacking magazine is placed on it. This arm then takes the prepared covering cardboard from a second stack, which contains not only operating instructions but also a section which can be pushed in and which prevents the full spool from unwinding in transit to the customer. The complete assembled cassette then passes through a pneumatic press which presses on the cover and presses in the protective cardboard. The third arm also requires only two degrees of freedom, so that half of the eight possible degrees of freedom provided by the control system are available to the central arm for the complicated threading operation (Figure 5.53).

Additional equipment:

Figure 5.54 shows the additional devices required, such as vibratory feeder, slides, magazines, belt conveyors and pneumatic press.

Reasons for the application:

- Simple setting due to recording of the programs either on punched or magnetic tape.

- Only one operator is required for monitoring the entire system, for topping up the magazines and vibratory feeders and for removing the assembled cassettes.

- In the event of incorrect assembly, a repetition is initiated automatically.

Task Assembly of typewriter ribbon cassettes	Example No. 7
Industrial robot used	Olivetti SIGMA/MTG
Cycle time (s)	6
Programming method	Teach-in method
Shifts/day	
Working area (mm)	1000×400×400
Number of arms	3
Number of degrees of freedom	3
Resetting time (h)	3 to 8 (depending on the cassette type)

Figure 5.53   General view of the SIGMA/MTG with 3 arms (Photo: Olivetti).

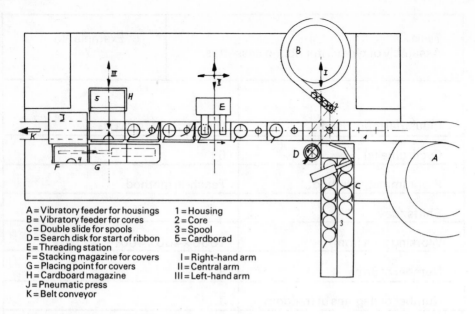

A = Vibratory feeder for housings   1 = Housing
B = Vibratory feeder for cores   2 = Core
C = Double slide for spools   3 = Spool
D = Search disk for start of ribbon   5 = Cardborad
E = Threading station
F = Stacking magazine for covers
G = Placing point for covers   I = Right-hand arm
H = Cardboard magazine   II = Central arm
J = Pneumatic press   III = Left-hand arm
K = Belt conveyor

Figure 5.54   Layout of the whole system.

flexibility is essential. This must be achieved by gripper changing systems or flexible grippers and flexible part storage devices. For more complex assembly operations the movements of several arms must in some cases be controlled in co-ordination. The loads, particularly in mating operations, cannot be determined exactly. The speeds must be high between the mating operations.

5.6.3   Industrial robots for workpiece handling

If industrial robots are used for workpiece handling they are equipped with grippers and have the task of bringing workpieces from a defined initial position to a defined end position. The path between the individual positions can be undefined. In most cases the units are point-to-point controlled and in principle differ from feeding devices only in their programmable movement cycle. The difficulties, and hence also the cost-generating factors, lie in the periphery, since sorting devices and grippers, for instance, do not have the same flexibility as the robot.

5.6.3.1   Handling on presses

Monotony and short cycle times are frequently the main characteristics of press work stations. The task of the human operator or the robot consists of feeding parts lying in a defined position into a press tool, or removing them from a press tool. The defined position is reached by previous orientating devices or magazines or the linking of several operations in which the orientated condition is maintained. Suction and magnetic grippers or mechanically actuated grippers are used as the gripping tools.

    Because of the long setting time for presses, the robots can also be reset during this period. Since the first workpiece detection systems to be used industrially can only be used for flat parts, the use of such sensors can initially be expected for the application with presses. In this case the industrial robots would have to be provided with the facility for external position stops.

206

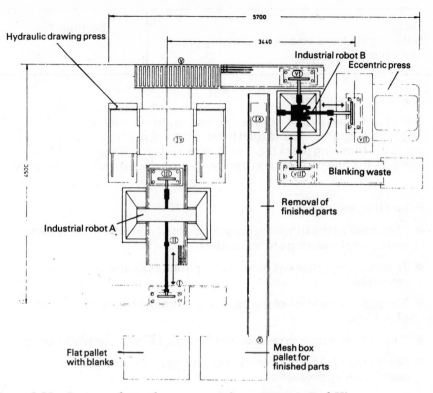

Figure 5.55   Layout of complete system (Photo: FIBRO GmbH).

*Figure 5.56:
Industrial robot A for
stacking and placing
blanks (3 movement
axes and 2 gripper
systems, 1 positioning
device). (Photo:
FIBRO GmbH).*

207

Task Handling of blanks on a deep-drawing press and a trimming press	Example No. 8

Statement of problems:

Blanks are to be fed into a drawing press, removed after the deep-drawing process and then fed to a trimming press. Before they reach the trimming press the parts are picked up and inserted. After removal, the finished part and the waste are to be deposited separately.

Boundary conditions:

● The stroke of the trimming press is so small that horizontal feeding and vertical lowering are impossible.

● The trimming press has no waste separation, i.e. the waste must be removed.

● The supply condition of the blank stack is a workpiece tolerance of $\pm$ 15 mm.

● It must be possible to reset the system for 18 different workpieces.

● Cycle time = 10 parts/minute (depending on the cycle of the deep-drawing press).

Solution:

The overall layout of the implemented solution is shown in Figure 5.55. Two industrial robots constructed on the module principle are required for this application, which means that the modules are identical but produce different units if suitably assembled.

Operating cycle:

Industrial robot A (Figure 5.56) removes the blanks from a stacking magazine (I) by means of a magnetic gripper and deposits them on a conveyor belt (II). There the blanks are positioned and taken up by a second gripper (III) and inserted in the drawing press (IV). The robot used here consists of a translational module fitted in a vertical position, which is mounted on a horizontal module. At the opposite end the blank is raised vertically by tilting the gripper by means of a tilting module.

After being ejected from the drawing press and positioned (IV) by means of a second conveyor belt (V), the blanks are picked up by robot B (Figure 5.62) and placed in the trimming press (VII). After the trimming process the workpiece and the waste are removed and deposited on a separating device. Finally the workpiece (VIII) separated from the waste is picked up by a third gripper and deposited on a roller conveyor (IX).

The industrial robot used here consists of a rotational module and two translational modules offset 90°.

Additional equipment:

● All the grippers are provided with initiators for workpiece monitoring. The grippers are designed for special profile strips for 18 different workpieces. The pick and place gripper of industrial robot A also performs the ejection function for the finished drawn part.

Reasons for the application:

Three operators were previously necessary for the stacking, placing and removal processes. Using the two units, this could be reduced to one-man operation, which was found to be particularly economic in view of the fluctuations in orders from the motor industry.

Task	Example No.
Handling of blanks on a deep-drawing press and a trimming press	8

Industrial robot used	FIBRO
Cycle time (s)	6
Workpiece weight (kg)	approx. 1
Shifts/day	2
Drive type	Hydraulic, pneumatic or electric
Control system	electronic (PTP)

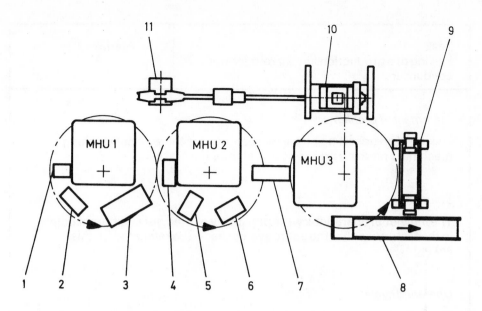

Figure 5.57   General layout of linking.

*Figure 5.58
Linking of a
production line
for refrigerator
condensers with 3
MHU (Photo: Robert
Bosch GmbH).*

Task Linking of a production line for refrigerator condensers	Example No. 9

Statement of problems:

Linking of an upsetting press, two bending machines, one punching machine, a piercing press and a straightening press.

Solution:

The individual operations were previously carried out at different points in the factory, so that the entire system had to be relocated for this linking.

Operating cycle:

The individual processing machines were installed in a circle around three industrial robots.

Robot 1 removes a pipe (Figure 5.57, item 1) from the upsetting press above the magazine, places the pipe in the first bending machine (item 2), removes a second pipe from the punching machine (item 3) and places it in the second bending machine (item 4). The pipe bent in the meantime in the first bending machine is now removed and placed in the punching machine.

Robot 2 has a similar operating cycle. Thus a pipe is removed from the second bending machine and initially stored (intermediate storage) (item 5). After a pipe is removed from the piercing press (item 6) and deposited in a straightening press (item 7), the stored pipe is collected and placed in the piercing machine.

Before Robot 2 can place "its" part in the straightening press, Robot 3 removes the pipe from the straightening press and places it in a pusher (item 10).

Task Linking of a prodution line for refrigerator condensers	Example No. 9

The pusher transfers the pipe to a press in which 43 cooling fins are progressively passed on to it.

After this operation, a workpiece with fins that have now been pressed on is removed from the parallel pusher (at item 10), fed to a press (item 9) for the cooling fins to be peened over, and deposited on a belt (item 8).

Reasons for the application:

- Rationalisation of the production process by combining different operations.
- Workplaces which previously necessitated extensive safety precautions (e.g. on presses) have been replaced by industrial robots.

Industrial robot used	Elektrolux MHU Senior
Cycle time (s)	30
Workpiece weight (kg)	up to approx. 4
Shifts/day	2 to 3
Drive system	pneumatic
Type of control system	electromechanical

5.6.3.2 Handling on forging machines

Forging machines, in most cases forging presses, impose the maximum physical demands on the human operator. Noise, heat, heavy weights, dirt, vibrations and accident risks make these workplaces particularly suitable for industrial robots. The extreme conditions, however, also impose stringent demands on an industrial robot, which is why many projects are doomed to failure from the very beginning, at least for cost reasons. Owing to the kind of forgings and the tool shapes, a high degree of positioning accuracy is not demanded from the industrial robot when feeding forging presses, but heavy workpieces must be manipulated at high speeds. Since the red hot parts are often subject to considerable changes in shape during processing, the grippers are in most cases very complicated and heavy. Forgings adhering the die necessitate additional forces. Kinematics (guides, drives) and control systems must be protected against large amounts of dirt, heat and vibration. Because of the fairly long setting times for the tools, long times can be expected here for the changeover.

Figure 5.59
General layout of the system (Photo: VFW-Fokker).

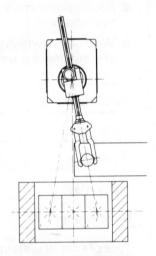

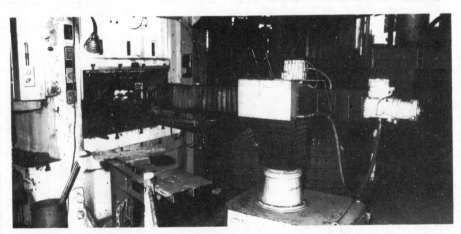

Figure 5.60   Feeding the forging press with the automatic transfer machine (Photo: VFW-Fokker).

214

Task	Example No.
Placing hot bar cut-offs in a forging press	10

Statement of problem:

Hot bar cut-offs up to 50 kg in weight must be placed in a forging press and transferred in turn to three stations for pre-upsetting, cupping and punching.

Solution:

The upright workpieces are placed by the industrial robot in the first station for pre-upsetting in the forging press (Figure 5.60).

To reduce the cycle time, the tongs either remain open in the press area during the press strokes, or keep the workpiece lightly clamped.

After the forging press, the ring is thrown on to a chute, whence it is fed to the following rolling mill.

Although the stations in the press are the same distances apart for all the workpieces, they must be reprogrammed each time (about once per shift), as the position of the gripper flange also changes with the workpiece diameter. A number of workpieces have to be turned through 180° between the second and third stations.

Additional equipment:

● Because of the high degree of deformation of the workpiece, a special tong design had to be developed for a wide clamping range.

● Differently shaped heat resistant gripper inserts can extend the working range to cover a wide workpiece spectrum.

Task	Example No.
Placing hot bar cut-offs in a forging press	10

Reasons for the application:

The handling of relatively large, heavy and red-hot workpieces by means of an industrial robot must be regarded as particularly advantageous from the point of view of humanisation.

Industrial robot used	VFW–Fokker
Cycle time (s)	
Workpiece weight (kg)	up to approx. 50
Shifts/day	2 to 3
Control system	electronic
Programming method	cross bar distribution

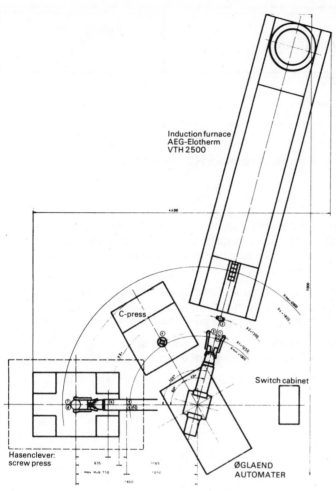

*Figure 5.61
Linking three forging
plants by an
industrial robot –
General layout
(Photo:
Zahnradfabrik
Friedrichshafen AG).*

Induction furnace
AEG-Elotherm
VTH 2500

C-press

Switch cabinet

Hasenclever:
screw press

ØGLAEND
AUTOMATER

Figure 5.62   Loading a forging press (Photo: Zahnradfabrik Friedrichshafen AG).

Task	Example No.
Linking of three forging plants	11

Statement of problems:

● An induction heating furnace, a pre-upsetting press and a percussion press are to be linked to an industrial robot.

● Frequent and therefore rapid resetting (5 to 10 times per week).

Solution:

A total of three complementary production units are linked by an industrial robot without peripheral equipment, with only one simple, robustly designed separating device being installed at the end of the induction heating system (Figure 5.61).

The loading of the pre-upsetting press and optionally one or two forging operations are carried out fully automatically by the industrial robot, whilst removal of the finished forging is carried out manually.

In a second automation stage, this process can be performed by a further robot, with simultaneous loading of a trimming press.

The required short cycle time of 8 s can be achieved by dividing the work in this way.

Task	Example No.
Linking three forging plants	11

Additional equipment:

● Adaptation of modified positions is effected through potentiometers with digital display and programming of the process control system, or by means of cross distribution or interchangeable diode matrices.

● Easily replaceable parts on the gripper permit rapid adaptation to modified tasks.

● In the event of faults, the robot can be moved gradually out of the danger area into the initial position.

Reasons for the application:

● Humanisation in the forging works – high physical strain occur under unfavourable environmental conditions due to heat and dirt and with short cycle times. Such workplaces can only be staffed with difficulty.

● High degree of utilisation, with simple operation.

Industrial robot used	ZF automatic handling machine T III XZ-1060
Cycle time s	8 to 20
Workpiece weight kg	up to approx. 35
Shifts/day	2 to 3
Programming method	Diode matrix or cross bar
Drive type	pneumatic
Number of manual axes	4

5.6.3.3 Handling on injection moulding and die casting machines

The material supply on pressure die casting machines takes place automatically. Industrial robots are used to remove the workpieces from the tool. The parts lie in a defined position, and in order to check that there are no workpiece residues left in the tool sensors are required, which can be integrated in the gripper. Tool spraying devices, which are in most cases fitted separately, can, under certain circumstances, be fitted to the gripper. The die casting is then often placed by the robot into a trimming device. No particular requirements are imposed on positioning accuracy, and the handling weights are relatively light due to the workpiece weights, which are technologically limited, but heavy where the grippers are heavy (e.g. with integrated spraying device). Because of the precisely defined cycle movement, the minimal change in position when the tools are changed and the long tool changing time, relatively simple unloading devices are required for unloading die casting machines. These are not programmable and are positioned by adjustable stops. To save space they are in most cases mounted on the die casting machine or designed as a gantry. The high proportion of programmable industrial robots can be justified by the relatively low peripheral expenditure. Industrial robots for unloading die casting machines must be insensitive to heat and dirt.

The operation of a flash removal injection moulding machine is basically the same as for a die casting machine. Here too, relatively simple single purpose machines designed for this application have been satisfactorily used in practice. However, industrial robots have been used with great success for linking with subsequent operations (flashing). Because of the short cycle times, higher speeds than for die casting are required.

The handling weights are generally lighter, but special conditions are often imposed on the gripper to avoid damage during gripping. When gripping the sprue, positioning in a flashing tool is problematic because the position of the part may vary in relation to the gripper because of insufficient hardening.

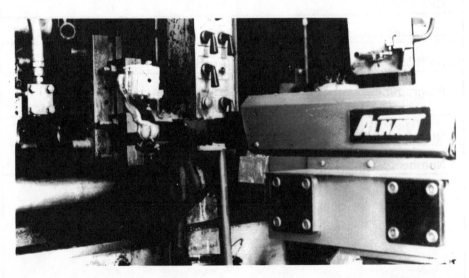

Figure 5.63   Handling on a pressure die casting machine with an ALMERT Re-5 (Photo: VFW-Fokker).

220

Task	Example No.
Workpiece removal on a pressure die casting machine	12

Statement of problems:

- Removal of workpieces up to 5 kg in weight from a die casting machine.

- Placing the removed workpieces in the cutting tool of a trimming press.

Solution:

The pneumatic industrial robot had five freely programmable axes of which the transverse movement, which lies at right angles to the horizontal axis and which is also horizontal, has been specially developed for applications on injection moulding machines, die casting machines and machine tools.

After the machine opens, the gripper enters it, grips the workpiece up to 5 kg in weight at the sprue, draws it at an adjustable speed in the direction of the longitudinal axis of the machine out of the tool without tilting it, and releases the machine for the next shot after leaving the working area.

This process only lasts a few second. The robot them places the workpiece, after 90° rotation of the gripper, in the cutting tool of a trimming press. After flashing the flash and sprue are again gripped and returned to the molten metal via a chute. Checking for completeness can then be interspersed between removal and release of the machine. Reprogramming is carried out about once a week.

Reasons for the application:

- Qualitatively better workpieces.

- More uniform thermal conditions in the casting machine.

- Shorter cycle times.

- Amortization within one year.

- Elimination of an inhuman workplace.

Task Workpiece removal on a pressure die casting machine	Example No. 12

Industrial robot used	ALMART Re-5 (VFW–Fokker)
Cycle time (s)	
Workpiece weight (kg)	up to approx. 5
Shifts/day	2 to 3
Drive type	pneumatic
Programming method	crossed bar distributor
Control system	electronic

Task	Example No.
Workpiece removal from injection moulding machine	13

Statement of problems:

- Removal of vacuum cleaner housing from an injection moulding machine and placing in a trimming press.

- Removal of vacuum cleaner housings from the trimming press and placing in a punching machine.

- Removal of the vacuum cleaner housings from the punching machine and deposition on a conveyor belt.

Solution:

The different production installations were grouped around an MHU industrial robot so that the individual stations to be approached were brought into the operating range of the industrial robot.

Operating cycle:

After automatic opening of the sliding door, the arm of the MHU moves into the tool area, grips the housing and withdraws. Whilst a new spraying cycle is started and is in progress, the MHU places the housings in a trimming press. After trimming, the housing is removed, first placed in a punching machine, removed again and deposited onto a conveyor belt.

Reasons for the application:

- The handling operations previously required were taken over by the MHU industrial robot. This means that only one operator is required to perform the inspection and monitoring functions, and it therefore permits multi-machine operation.

- Uniform quality.

Task Workpiece removal from an injection moulding machine	Example No. 13
Industrial robot used	MHU (Robert Bosch GmbH)
Cycle time (s)	
Workpiece weight (kg)	approx. 0.5
Shifts/day	2 to 3
Drive type	pneumatic
Programming method	
Number of axes	4

Figure 5.64   Gripper with vacuum cleaner housing (Photo: Robert Bosch GmbH).

5.6.3.4   Workpiece handling on machine tools

The use of industrial robots for the loading and unloading of machine tools is numerically small. The reasons for this lie mainly in the multitude of secondary functions which must be performed in addition to the handling function. These include, in particular, orientating functions, inspection and auxiliary functions. In this field the method of "workpiece thrown into the box" is still widespread, which necessitates re-orientation at each workplace. The inspection functions carried out include workpiece inspections and machine and tool monitoring. Auxiliary functions such as chip removal cause special difficulties, thus the successful use of an industrial robot is only possible after the problems relating to the secondary functions have been solved. The use of industrial robots on machine tools is there-fore often found in flexible production systems where the functions mentioned are largely automated. Stringent requirements are imposed on the robot with regard to positioning accuracy and speed (very fast and very slow). The handling weights are mostly great because complicated multiple grippers are required due to the minimum workpiece changing times that are necessary. In flexible production systems, flexible grippers can be used under certain circumstances, and these also represent a considerable load for the industrial robot. As the setting time for machine tools is often very short, robot must be quickly reprogrammable for these purposes. As the industrial robot is generally not in use during a tool change, its use for the tool change is conceivable, but there too the robot must operate quickly and accurately.

225

Statement of problem:

Loading of a hobbing machine with workpieces from a portable cyclic storage unit.

Solution:

Figure 5.65 shows the arrangement between the industrial robot, the portable cyclic storage unit and the hobbing machine. The operating cycle of the machine, the robot and the storage unit is synchronised by the control system of the robot. In the storage unit the workpieces are deposited already orientated, retain this arrangement and are offered to the robot for removal.

Additional equipment:

Portable cyclic storage unit (Figure 5.66).

An increase in flexibility is achieved by the use of these storage units wherever the use of fully automatic conveyor systems is not possible for reasons of space or due to the type of production. Single pallets, interchangeable for different workpiece dimensions, are connected endlessly to each other by a conveyor chain, and are driven by a geared motor with an electro-magnetic coupling at the cyclic speed.

Advantages of these cyclic storage units are as follows:

● Standardisation of storage and conveyor system for any number of replaceable and complementary production devices.

● Optional for manual loading and unloading or suitable for use with an automatic handling device.

Task Workpiece handling on a hobbing machine machine	Example No. 14

- Rapid adaptation to the widest possible range, with different workpiece dimensions, without additional adaptors
- Sufficiently accurate workpiece positioning and orientation definition
- High degree of operating reliability
- High degree of accident safety
- Simple, economical design

However, "flexible production systems" can also be formed with such storage units. Figure 5.67 shows a possible arrangement.

Reasons for the application:

- Relieving of operating personnel of adherence to cycle
- Independence of firm-specific automation devices
- No rigid linking
- Optional assignment of the cycle storage device to the production units

Industrial robot used	ZF-automatic handling machine S II CXZA-825
Cycle times (s)	8
Workpiece weight (kg)	up to 5
Workpiece diameter (mm)	50 to 200
Shifts/day	
Resetting frequency	2 to 3 times a week
Drive	pneumatic
Number of manual axes	4

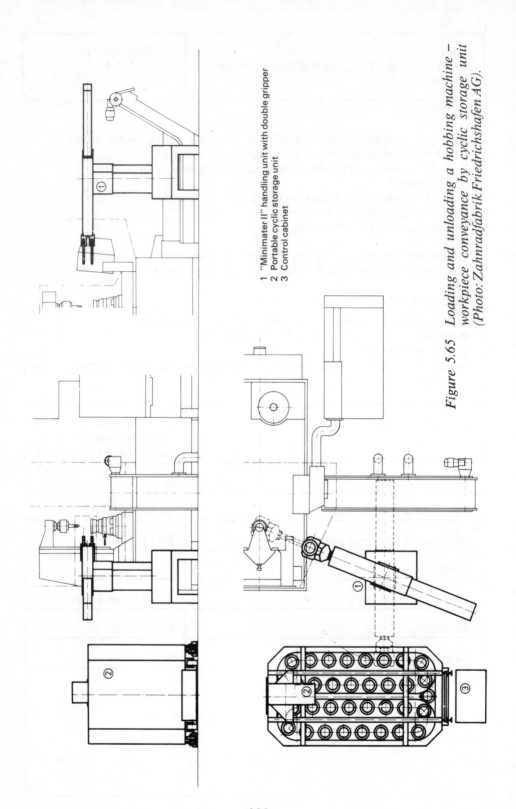

1 "Minimater II" handling unit with double gripper
2 Portable cyclic storage unit
3 Control cabinet

Figure 5.65 Loading and unloading a hobbing machine – workpiece conveyance by cyclic storage unit (Photo: Zahnradfabrik Friedrichshafen AG).

228

Figure 5.66   Portable cyclic storage unit (Photo: Zahnradfabrik AG).

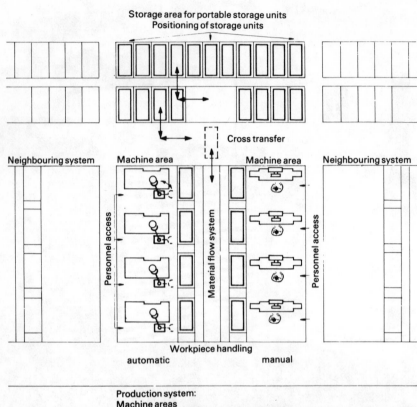

Storage area for portable storage units
Positioning of storage units

Cross transfer

Neighbouring system   Machine area   Machine area   Neighbouring system

Personnel access   Material flow system   Personnel access

Workpiece handling
automatic   manual

Production system:
Machine areas
Positioning points for portable storage units
Material flow system

*Figure 5.67   Flexible production system using portable cyclic storage units
suitable for automatic and manual machine loading.*

230

Task Workpiece handling for trimming an injection moulded part	Example No. 15

Statement of problem:

Trimming a container housing manufactured by die casting with the aid of drilling and milling units.

Solution:

The workpieces to be trimmed are placed manually on a slide in the correct position on which they slide through gravity against an end stop in a defined position.

There the workpieces are picked up by a robot with a special gripper (Figure 5.68) and taken, already positioned, to the work station provided (Figure 5.69), with simultaneous and additional positional orientation. The work stations are arranged in a semi-circle around the robot. On completion of the trimming process, the workpiece is deposited on another slide.

Additional equipment:

● Two slides for loading and removing the workpieces.

● The robot and the individual work stations were installed in an enclosed cabin; all the necessary programming work can be carried out from the outside.

Reasons for the application:

● Trimming of the contactor housing is necessary for the subsequent function and for asembly, and must therefore be carried out reliably and accurately. Manual trimming was time-consuming and subject to error. Through the use of an industrial robot uniform quality could therefore be achieved and the error rate reduced.

Task	Example No.
Workpiece handling fo trimming an injection moulded part	15

● The previously monotonous operating process, combined with unfavourable environmental conditions (incidence of dust) meant the elimination of an inhuman workplace through the use of an industrial robot.

Industrial robot used	ASEA (b 6)
Cycle time (s)	62 (48 s of which is trimming time)
Shifts/day	2 to 3
Output/shift (pieces)	approx. 450
Drive type	electrical
Programming method	teach-in method

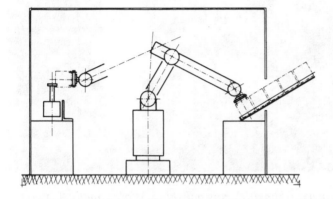

Figure 5.68
Layout of the
whole system.

Figure 5.69   Handling of an injection moulded part during the trimming process
            in several work stations with a b6 industrial robot (Photo: ASEA).

5.6.3.5 Industrial robots for guiding components

An industrial robot occupies a special position when a component is held or guided along a defined path during the machining process. This is mostly the case when an industrial robot is used for a machining process and when the component to be machined is ligher than the tool.

Figure 5.70   Handling of a washbasin during polishing (Works photo: ASEA).

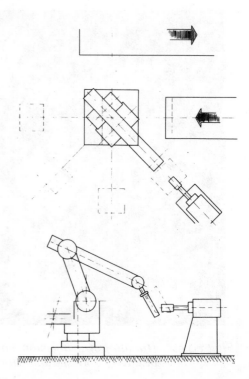

Figure 5.71
General layout of the system
(Photo: ASEA).

Task	Example No.
Polishing a high-grade steel washbasin	16

Statement of problem:

High-grade steel washbasins must be finish-ground and polished after deep-drawing and stress relief annealing.

This process has previously been carried out by workers who emery and polish the basins at polishing units.

Solution:

The washbasins arriving on conveyor belts are picked up by an industrial robot with a special gripper and placed in position in front of different polishing stands (Figure 5.70), each of which polishes an allocated area. The polishers are arranged in a circle around the robot (Figure 5.71). On completion of the polishing process, the washbasins are deposited on another conveyor belt.

Additional equipment:

● Two conveyor belts for feeding and removal of the workpieces respectively.

● Safety enclosure – on entering this enclosure, which encompasses the entire workspace of the robot and the different work stations, the plant is stopped.

Reasons for the application:

● Elimination of an inhuman workplace. This workplace was difficult to man because of heavy weights to be handled (up to 15 kg) and need for considerable strength (contact force of 30 to 40 kg) and because of the considerable amount of dust produced.

Task Polishing a high-grade steel washbasin	Example No. 16

- Polishing of ten different washbasins, which vary in size and shape, was made possible by an industrial robot.
- Rapid program change due to reading in of the appropriate cassette.

Industrial robot used	ASEA b60
Workpiece weight (kg)	up to 15
Cycle time (s)	120 to 180, according to basin shape
Shifts/day	2
Programming method	Teach-in method
Drive	electric

5.6.3.6 Miscellaneous

Numerous applications for industrial robots which cannot be included in the above-mentioned sections, e.g. the interlinking of different production units, are often simple transfer processes but processes which are in some cases made difficult by the fact that the production process is continuous and necessitates the picking-up or depositing of workpieces into moving workpiece fixtures. Other examples of industrial robot applications are paletting processes, e.g. for stacking workpieces in a certain order.

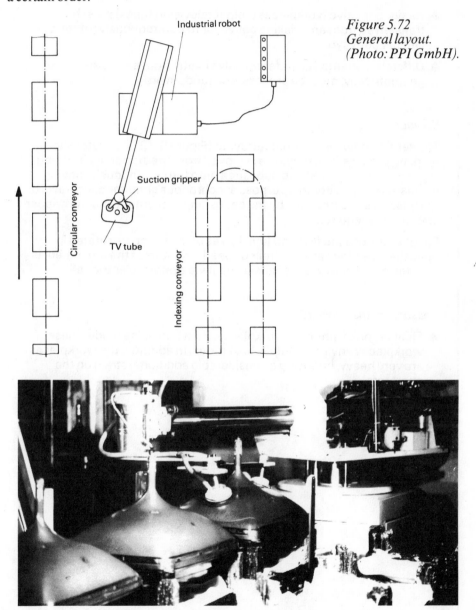

Figure 5.72
General layout.
(Photo: PPI GmbH).

Figure 5.73   Removal of television tubes from the indexing conveyor (Photo: PPI GmbH).

237

Task	Example No.
Transfer of colour television tubes	17

Statement of problems:

● Heavy, sensitive workpieces (colour television tubes) must be transferred from an indexing conveyor to a continuously rotating circular conveyor.

● Different variants, but differing only slightly from each other geometrically, must be processed in random order.

Solution:

The set-up shown in the general layout (Figure 5.72) illustrates the working process. The tubes are removed from the stationary indexing conveyor (Figure 5.73) and deposited onto the continuously moving circular conveyor. For this purpose synchronous simultaneous travel of the gripper, superimposed by the depositing movement and the gripper operation, is required.

For this purpose the first and fourth axes of the robot are rotated in opposite directions at a controlled speed. The start of the movement is initiated by the moving suspension gear via a photo-electric cell.

Reasons for the application:

● Elimination of inhuman workplaces. Heat and noise render these workplaces environmentally hazardous. In addition, the workpieces are very heavy, imposing a considerable additional strain on the operators.

Task	Example No.
Transfer of colour television tubes	17

Industrial robot used	PM 12 (PPI GmbH)
Cycle time (s)	
Workpiece weight (kg)	up to approx. 18
Shifts/day	3
Programming method	– direct computer control – by input with data carriers – approaching the required positions at creeping speed and in single steps
Control system	Multi-point and continuous path control

Task	Example No.
Handling of rear axles between the measuring and welding device	18

Statement of problem:

During the production of the rear axle it must be placed in a welding fixture for welding and removed again after the welding operation.

Solution:

A robot picks up a rear axle arriving on a conveyor belt in a defined position, using a special gripper, and places it in the welding fixture.

After the welding operation, the rear axle is again removed by the same robot and deposited onto a chute (Figure 5.74).

Additional equipment:

● Rear axle positioning device for defined orientation.

● Gravity chute to the pallet conveyor belt.

● Protective enclosure with a safety door.

● Changing of the clamps in the welding fixture.

Reasons for the application:

● Physically heavy work is taken over by an industrial robot (previously two men were required to do this work).

Task Handling of rear axles between measuring and welding device	Example No. 18

Industrial robot used	Tubular unit (VW AG)
Workpiece weight (kg)	17
Cycle time (s)	28.5
Shifts/day	2
Programming method	Teach-in method
Drive	Electromechanical
Number of axes	3 primary, 3 manual axes

Statement of problems:

Loading and unloading of an hydraulic press for processing univeral shafts. The workpieces are positioned vertically at (A) and must be loaded into the tool (W) in the same position. After the operation the workpieces are removed from the press and suspended vertically in a circular magazine (M).

The tool is only accessible from the front of the press. The press itself is part of a production line which operates at a rate of 480 parts per hour in 2 shifts.

Solution:

The existing production line was interlinked by the installation of 2 FELSOMAT minirobots, model FE 30 (R_1 and R_2), as shown in the diagrammatic sketch (Figure 5.75). Figure 5.76 shows a section from the complete system.

The press is loaded by minirobot R_1, which is equipped with a rotary gripper.

Due to predetermined workpiece positioning and the design of the tool, a rotational movement of the gripper tongs about a vertical axis is required. For unloading, a second minirobot R_2 with a telescopic gripper arm is used. The telescopic movement of the gripper arm enables the workpiece to be removed from the tool and placed in a circular magazine (M).

Task	Example No.
Machining of universal shafts	19

Additional equipment:

The correct positioning of the workpieces is done by the receiving station (A), equipped with an indexing device.

The removing equipment was mounted on the press with an articulated bracket (K), which provided optimum adjusting facilities for the robot between the press and the circular magazine.

The entire interlinked system is controlled by an electronic/electro-mechanical control unit.

To prevent accidents, the system is protected by a grill (S).

Reasons for the application:

● Since approx. 10 tonnes of workpieces have to be moved each shift, it was absolutely necessary to relieve the operator and introduce a mechanical handling system.

● The implemented solution is economical as the interlinking expenditure can be recovered in approximately 18 months.

Industrial robot used	FELSOMAT FE 30
Cycle time (s)	7.5
Workpiece weight (kg)	1.8
Shifts/day	2

Figure 5.74   General view of handling rear axles (Photo: VW AG).

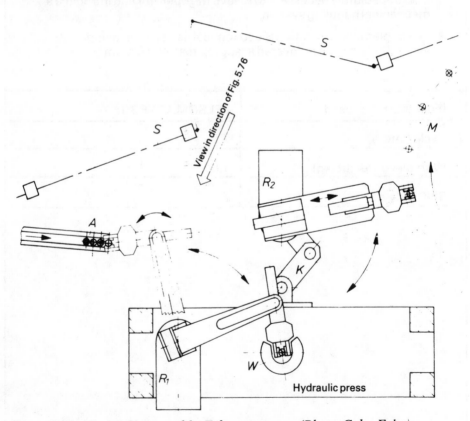

Figure 5.75   General layout of the Felsomat system (Photo: Gebr. Felss).

244

Figure 5.76   Part of the complete system (Photo: Gebr. Felss).

5.7 Work safety

5.7.1 Problem and task of work safety

The problem of working safety may be tackled from various points of view. The demand of employees of their representative for a more humane arrangement of working processes and workplaces is opposed by the aim of the employers and company owners to achieve the greatest output with the means available. At first sight there appears to be an unbridgeable conflict of objectives between the interests of the employee and those of the company in the field of working safety, where any advance to achieving one of these objectives only being possible at the expense of the other.

Only when the ways of looking at this problem are discriminating is it apparent that working safety measures can indeed satisfy both objectives, for it is

- the individual and vital interest of everyone to remain physically and mentally healthy.
- the social interest of the companies to protect the employee from injury,
- the economic interest of the company owners to avoid direct or indirect costs resulting from personal injuries,
- the legal and social interest of the government to enact laws and regulations applicable to working safety, and to ensure that they are observed,
- the interest of all those participating in the national economy to maintain the health and work capacity of everyone and to avoid the costs resulting from personal injuries (5.22).

One of the tasks of accident research is therefore to analyse the tecnical factors and the factors dependent on human behaviour which (may) lead to an accident, and to develop effective methods of accident prevention. This goal can be achieved in two different, but complementary ways.

245

- by a forward looking, analytical determination or
- by a retrospective determination of the accident risk derived from the study of past accidents.

In principle it would be preferable if the first method were chosen, because here lessons do not first have to be learnt from accidents that have already happened. The main problem lies on the one hand in the difficulty of finding objective criteria determining the degree of risk dependent on technical and personnel-related conditions at the workplace, and on the other in reducing the expenditure involved to practicable dimensions.

Thus even in the future it will not be possible to dispense with the task of determining accident risk from accidents that have already occurred. The practice still commonly used today of limiting accident prevention to the taking of measures which only avoid a repetition of the accident is not sufficient. This applies all the more with the increasing complexity of the conditions which led to the accident (5.23).

5.7.2 Automation and work safety

As easy as it is to realise that the living and working conditions of the individual are decisvely influenced by increasing technological development and automation of formerly manual activities, it is just as difficult to make any valid statement on the quality of the overall changes to be expected. Though technical innovations are the direct cause of changes as such, their consequences are largely determined by the "social ground" onto which they fall and the paths along which they are steered.

This also applies to automation which, if justified solely from the point of view of short-term cost saving, may give rise to a number of socially undesirable side effects, which, if also viewed from other objectives, may, however, provide a predominantly positive balance. One of these objectives is the improvement in working safety in order to reach a position, as defined in the relevant literature, "in which the human being is protected against risks which may produce accidents, occupational diseases and other occupational physical injuries". Whilst all the activities which contribute to producing a product are still carried out manually in officially designated manual work, more and more technical devices are now being used for this purpose at higher production stages, until finally, in the last stage, fully automated production is achieved by the mechanisation of the handling activities.

From the point of view of working safety, this process must be regarded as a positive development when the manual work created an unduly high risk to the human operator, a risk which is eliminated by the use of a technical aid, in the sense that the risk itself is removed or the human operator can be kept away from the danger area.

5.7.2.1 Automation of dangerous handling operations

Accidents occurring in handling operations are probably one of the most frequent types of accidents anywhere, as at first glance at the statistics on the frequency of injured parts of the body shows. Every second accident (amounting to 1 million in 1974) involves injuries to the hands and arms with which handling is, after all, carried out.

At the top of the accident frequency list lies pressure die casting. What is striking here is the high proportion of "burns caused by molten metal" (approx. 75% of all accidents), whilst other causes are only of secondary importance (Figure 5.77).

The fact that this result is to be expected is apparent if the normal working process on a pressure die casting machine is broken down into individual

246

Accidents on pressure die casting machines in the years 1969-1973

Year	1969	1970	1971	1972	1973
Number of accidents reported	–	340		233	
of which investigated	168	245	216	186	204
Frequency of the different causes of accidents (%):					
Burns from molten metal	71	72		77	79
Metal escaping from the mould	28	30		27	19
Burst casting residues	14	17		23	27
Spilling of metal when pouring into the pressure chamber	13	11		11	14
Other burns caused by molten metal	16	14		16	19
Other causes %	29	28		23	21
Accidents due to squeezing points including closing movement of mould	3.5	3		5	4
Accidents during setting	6.5	9		7	10
Accidents due to burning hydraulic fluids	2.5	1			
Other causes	16.5	15		11	7

Figure 5.77   Frequency of different causes of accidents in the pressure foundry according to (5.24).

Operation	Possible accident risks	Stress conditions	Remedies
(1) Closing mould	Bruising, tearing and severing due to closing mould, injuries due to metal splinters if the mould is damaged		Mechanical protective devices such as protective doors or automatic handling equipment
(2) Scooping and feeding metal ●	Burns due to metal splashing and spilt metal, contusions from ladle	Heavy physical work, heat, fumes from burnt residues such as piston lubricant	Automatic scoop, suction
(3) Forcing in metal	Burns due to metal escaping from the mould	Noise	Enclosing the parting line of the mould on all sides
(4) Opening mould	Burns due to bursting casting residue	Heat, fumes from burnt mould spraying agent	Protective clothing, protective helmet and face protection, suction or automatic handling equipment
(5) Removing workpiece ●	Burns due to bursting casting residue and, on the mould, bruises, contusions and burns due to falling workpiece, catching on the tongs	Heat, fumes, heavy physical work depending on workpiece weight	Protective clothing, safety shoes, or automatic handling equipment
(6) Inspecting workpiece (visual inspection) ◑	Burns due to bursting casting residue, bruises, contusions and burns due to falling workpiece, catching on the tongs	Heat, fumes, heavy physical work depending on workpiece weight	Automatic inspection device together with automatic handling equipment, protective clothing
(7) Trimming ◑			
(8) Spraying of mould ●	If the control system fails, bruising, contusions, severing due to closing mould, burns from the mould	Heat, fumes, humidity, noise	Automatic mould spraying by mould spraying unit or handling unit
(9) Placing inserts in mould ●	If the control system fails, bruising, contusions, severing due to closing mould, contusions due to catching on the mould, burns from the mould	Heat, fumes	Automatic placing by handling equipment

● Handling operation
◑ Includes handling operation

Figure 5.78   Relationship between operating process, accident risks and stress conditions in pressure casting.

248

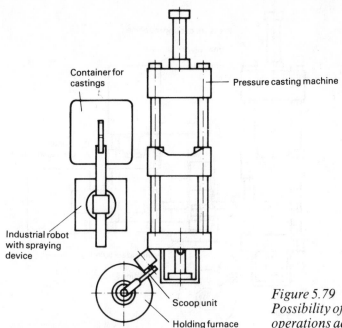

Figure 5.79
Possibility of automating handling
operations according to solution A.

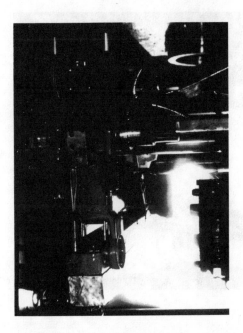

Figure 5.80
Example of implementation according
to solution A (Photo: ASEA).

operations, and the risks involved in each operation are considered separately
(Columns 1 and 2 in Figure 5.78).

Two things are remarkable here:
1. Five of the nine operations are handling operations.
2. Seven of the nine operations involve the risk of burning.

The risks involved in operations 1, 2 and 4, in which only a switching process is

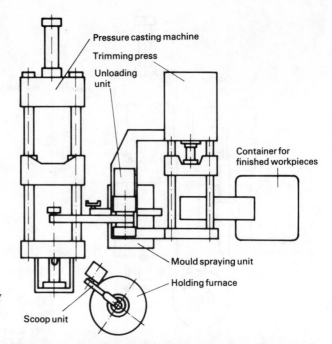

Figure 5.81
Possibility of automating
handling operations
according to solution B.

Pressure casting machine

Trimming press

Unloading
unit

Container for
finished workpieces

Mould spraying unit

Holding furnace

Scoop unit

Figure 5.82
Example of
implementation
according to
solution B.

initiated, or the stopping of a machine must be awaited, are primarily associated with handling tasks.

From the point of view of proportions of time, the breakdown will be roughly as follows:

2/3 handling operations

1/3 other operations

Since handling work predominates from the point of view of both the time and

250

Persons at risk	Prevented accident risks and stress conditions	Additional accident risks and strains/stress conditions if no additional safety devices are used
Operator (work paced by operating 8 cycle)	Burns from molten metal during scooping and molten metal escaping from the mould	
	Burns due to bursting casting residue, dropping of workpiece on the mould	
	Bruises and severance of limbs due to closing mould	
	Bruises and contusions due to falling workpiece or tongs	
	Noise, heat, fumes, humidity, heavy physical work	
Setting up repair personnel, monitoring personnel		Bruises and contusions from automatic scoop unit, automatic removal unit and automatic mould spraying unit
		Burns on the automatic scoop unit
Persons not working at the workplace (unauthorised and inquisitive persons)		Bruising and severance of limbs due to closing mould (no longer supervised by operator), bruises and contusions from automatic scoop unit, automatic mould spraying unit
		Burns on automatic scoop unit

Figure 5.83   Change in working safety after automating handling.

number of operations, the machine operator stays with the die casting machine throughout the operating cycle.

As far as working safety is concerned, this means:

Even during the operations in which his presence is not actually required in the risk area, the operator is subject to an accident risk. This also shows up in the accident statistics if we look, for example, at the line entitled "Metal escaping from the mould". Over one quarter of all accidents were caused by this. The fact that the figures from 1973 onwards are somewhat lower is due to the increasing use of moving guards from this point onwards. However, this measure still had no effect on the proportion of other accident causes. Since the main risks are directly related to handling operations – i.e. to the scooping up and pouring of the molten metal, the removal of the casting and (to a small extent) the spraying of the mould, suitable working safety measures should be taken here. One possibility, which will be demonstrated in the following by two examples, consists in removing the human operator from the immediate risk area and arranging for his work to be carried out by handling equipment. In the first solution (Figures 5.79 and 5.80), the molten metal is fed in by means of a scoop unit, whilst an industrial robot first takes the casting from the mould, places it in a container and then applies the mould release agent to the mould.

The other variant (Figures 5.81 and 5.82) consists of three single units: a scoop unit, an unloading unit and a spraying unit. By contrast to the first solution, the casting is also fed into a press for deflashing.

Which solution is considered to be more suitable in an individual case depends upon several different parameters, such as space available, cycle time, interlinking system, the required motion cycles for removal and spraying.

As far as working safety is concerned, however, both solutions must be regarded as equivalent. The success which can be achieved by such measures is indicated in the accident statistics for the industrial field of "Pressure casting works" for one company, where in 1975 the proportion of "burns" ws only 27% and "other cuases" 73%. Here about 55% of all pressure casting machines are equipped with automatic scoop devices, and about 20% with automatic or semi-automatic work-piece removal equipment.

The following overall picture will be seen in pressure casting works with regards to working safety after automating the handling operations (Figure 5.83).

For the former operator, who now only carries out monitoring work (not dependent on machine cycle), the situation has improved considerably. However, there are new additional risks which are produced by the handling equipment itself.

Here aspects of working safety must be considered already in the planning phase, in order to prevent lessons being learnt solely from accidents that have already occurred.

5.7.3 New accident risks which occur through the use of industrial robots

5.7.3.1 Accident risks peculiar to industrial robots

Accident risks due to the use of industrial robots are mainly the result of motions with large kinetic energy (high speed travel, movement of heavy weights). Of course these risks exist also in other conventional machines and systems, but they have a less serious effect, as illustrated in the comparison in Figure 5.84.

Industrial robot	Conventional machines
Simultaneous movement in several (5 to 6) axes	In most cases only simultaneous movement of few (1 to 2) axes
Free programmability of the speed of of each individual axis	Fixed predetermined speed
Free programmability of the direction of movement of each individual axis	Fixed predetermined cycle of movement
Very large workspace in relation to volume of equipment	In most cases workspace smaller than machine volume
Workspace overlaps the standing space of other machines, parts of buildings, etc.	

Figure 5.84   Comparison of the specific characteristics of industrial robots and conventional machines which affect working safety.

Because of the possibility of travelling along large number of axes simultaneously and selecting the positions to be approached and the travel speed for each axis individually. It is not possible even under normal conditions, for an outsider to anticipate the next movement. Moreover, in the event of a fault (e.g. in the displacement measuring system or in the speed monitoring system), completely unforeseeable movements may occur at an undefined speed within the (kinematically possible) workspace. Since, by contrast with conventional machines, where in most cases movements take place inside the machine, the size and shape of the risk area in the case of the industrial robot, i.e. the size and shapre of the working area or maximum workspace, cannot be detected without difficulty, precautions must be taken to screen off and protect this area. Another safety problem which is, of course, not specific to industrial robots but concerns all automatic systems, is added, namely the greatly reduced reliability of the whole system when different machines and systems are interlinked. Here malfunctions of individual subsystems may not only lead to the incorrect transmission of signals (i.e. starting or stopping the industrial robot at the wrong times), but may also lead to the incorrect transmission of displacement and speed information (i.e. the industrial robot travels along incorrect paths into incorrect positions).

5.7.3.2   Personnel groups at risk
Safety problems are encountered for different groups of personnel, according to the nature of the operation of the industrial robot.

Automatic operation
In automatic operation it is mainly
– the operating and monitoring personnel
– the personnel at neighbouring workplaces and
– inquisitive persons,
who are particularly at risk.

Here measures must be taken, by means of fixed barriers, covers, protected areas, etc., to ensure the prevention of the danger area being entered or broken into.

This means that the operating process should be organised in principle so that no spatial and temporal human robot interaction can occur either in space or in time. However, if entry into the working area of a robot is absolutely necessary (e.g. to load workpieces), a protective device must be provided which cannot be opened until completion of the hazardous motion and which must be re-closed and locked before this motion can be repeated.

Programming and setting-up
Programmers and setters are at risk when they have to enter the danger area to carry out their work. As it is often not possible to employ protective devices here, it is all the more important to design the overall system so that safety measures can only deliberately be removed for a specific purpose by persons authorised to do so.

The EMERGENCY OFF circuit, however, must remain operational in all cases. For setting purposes, a separate (key) switch should have to be operated, which only permits the movement of the industrial robot at creeping speed and/or by inching and which also prevents the whole system from being accidentally put into operation whilst the setter is still working in the danger area.

In addition care should be taken in the planning to place the robot and the other machines so that no pinching or shearing positions can occur during the operation in order to reduce the seriousness of any injuries incurred. Adjacent walls, pillars and fixed guards (fences) must also be included in the considerations.

Repair and maintenance
Those who have to determine the cause of a breakdown (if they have to enter the danger area for this purpose) are primarily at risk, as are the personnel who have to remain in the danger area in order to remove the fault or carry out the repair. Indeed most of this work can be carried out when the installation is stationary, but for work in which this is not possible the same safety systems should be in operation as for automatic operation and during setting-up. However, as maintenance personnel by contrast to other persons at risk, are able to put all the safety devices out of operation, whether intentionally or accidentally, care should be taken firstly that the whole system is designed so that it is clearly laid out and it has definite interfaces between its individual units, and secondly that measures must be taken (fault diagnosis systems, repair schedules, special tools, preventive maintenance, etc.), which facilitate the work of the maintenance personnel and thereby make it safe.

5.7.4 Safety measures and regulations for industrial robots
Since there are at present no (binding) accident prevention regulations for the construction and operation of industrial robots, some suggestions can be made here for safety measures for the application of industrial robots on the basis of the existing general regulations and recommendations (DIN 31000 and 31001, VDE 0113, Gen. Accident Prevention Regulations), and applications so far implemented in practice. The draft of a VDI Guideline (VDI 2853), which deals with the problem of safety of industrial robots and other handling equipment, is only availble at present as a so-called "Green Paper" (June 1979).

5.7.4.1 Design for safety of industrial robots
Industrial robots should – similar to DIN 31000 and the Arbeitsmittelgesetz (Production Equipment Act), Art 3 – be designed so that risks are excluded wherever possible, exhausting all technically and economically justifiable possibilities. This relates in particular to the following characteristics of this equipment:

Mechnical strength (see also DIN 31000).
Industrial robots should be provided with sufficient mechanical strength to ensure

that within certain limits, either a safe-live behaviour or limited failure (fail-safe behaviour) is achieved, even in the case of overloading, material defects, wear and other unfavourable physical and chemical influences.

Kinematics (see also DIN 31000)
Industrial robots should be designed so that moving parts which represent a risk are not accessible and cannot be touched. Since this requirement opposes the functions of the industrial robot, which is to perform movements (handling operations), provisions must be made, by technical safety aids and measures, to ensure that personnel are not endangered (see Section 5.7.3.2 "Automatic operation", "Programming and installation operation", and "Repair and Maintenance").

In the external design of industrial robots efforts should be made, where technically possible, to avoid points and sharp edges and corners, particularly on the furthest projecting axes.

Control (see also DIN 31000 and VDE 0113)
The switch and control mechanisms should be designed and arranged so that undesirable or unintentional operation is prevented.

In the case of automatic switching and control processes, the course of the predetermined sequence of functions must be guaranteed by a suitable interlocking system.

The control process, if it is electrical, must be designed to comply with VDE 0113. The same applies to other types of control system.

Industrial robot control systems must be able to be stopped quickly, easily and without risk by means of EMERGENCY OFF devices. The EMERGENCY OFF switches must be provided in sufficient quantity and must not permit any dangerous movements due to residual energy or over-running after actuation. Restarting after EMERGENCY OFF actuation must only be possible after manual release.

Switching to "Setting" operation, where some of the protective measures must be neutralised under certain circumstances, must only be possible by means of a key switch. In this case, however, care must be taken to ensure that in all cases the EMERGENCY OFF circuit remains in operation and that a movement of the industrial robot is only possible at creeping speed and/or inching operation. Control faults must not make both these protective measures ineffective.

Drive (see also DIN 31000 and VDE 0113)
Particular care must be taken to ensure that:
– in the event of an energy failure (or failure of the drive system) no dangerous movements can take place (collapse or overrunning of individual robot axes) and
– when the energy is restored, no automatic starting of the equipment is possible.

Grippers
The design of the grippers should also avoid sharp edges and corners wherever possible.

In addition, it should be ascertained in each case whether it is better retained due to the form, from the point of view of working safety, to ensure in the event of energy failure that the workpiece remains in the gripper (or by means of a spring force), or that the workpiece (e.g. hot part) is released under controlled conditions.

When designing the gripper consideration must also be given to the fact that the loading capacity of the robot specified by the manufacturer does not generally include the gripper weight.

Working safety considerations in the layout of workplaces with industrial robots

The entire system should be clearly arranged with the individual machines interlinked so that well-defined, easily recognisable interfaces are produced. It should be possible to switch the different machines and units on and off individually.

In the planning phase, consideration should be given to whether there is any possibility of carrying out maintenance and repair on the robot oustide the plant, the plant continuing to run in this case using a replacement unit.

In order to facilitate maintenance and repair work and thereby make it safer, the machines and units should be installed so that they are easily accessible. This applies particularly to those places where wearing parts or other components subject to frequent breakdowns are located.

Furthermore, it should be determined whether, and in which places, aids can be provided for better fault diagnosis, e.g. in the form of control lamps, since a number of accidents happen to maintanance and repair personnel when looking for faults.

Protective devices should be integrated in the system and it should only be possible to switch them off deliberately (e.g. by actuating a key switch) deliberately in a limited, easily ???? area by authorised persons.

When planning the operating cycle, care must be taken to ensure that manual operations are not allowed to be carried out together with automated operations at the same place. If this is unavoidable, protective measures (controlled transfer windows, two-handed switches, floor contact switches, photo-electric protective areas and similar devices) must be provided to reliably prevent human/industrial robot interaction.

The entire danger area should be surrounded by a protective fence the accesses to which are interlocked to VDI Guidelines 3231 (see also DIN 31000, 31001).

5.7.4.3 Work safty measures in the operation of industrial robots

The degree of working safety that can be achieved at workplaces using robots in particular is determined essentially when designing machines and equipment and when planning the layout.

If it can be assumed that all the possibilities have been exhausted in this respect, within a predetermined framework, the operation can be concentrated on the following points as part of its contribution to working safety (these are not specific to industrial robots but apply generally to technical installations):

Weak points in working safety which may not become apparent until some time after the system has been commissioned, must be reported to the competent departments (Safety, Works Planning). This will not only ensure that these weak points are eliminated but that these faults can be avoided in future.

Planned preventive maintenance (maintenance plans, working plans for repair, special tools) reduces not only unplanned machine down times, but because of the improved working conditions for the maintenance fitter also contributes to increasing working safety.

Regular instruction and training of operating, setting and maintenance personnel, and motivation to safe working, are further possibilities of reducing the accident figures.

6. Development trends

6.1 Sensor technology

Many workplaces in industrial production require not only handling functions, but at the same time also testing and monitoring functions. In these fields, the human being, with his sensory capabilities and learning ability, is far superior to a machine, since he is able to adapt to varying boundary conditions. In order to obtain a similar adaptability with robots they will have to be equipped with appropriate sensors with rapid and reliable signal processing, which are suitable for the direct control of an industrial robot.

Though the scope of industrial robots is extended considerably with the use of sensors for workpiece position measurement and identification, the decisive step towards detecting the environmental area is not yet possible with the present sensors. This step would be required if the problems in production are to be overcome.

In many industrial fields, parts of production processes are automated, but because of his highly developed sensory capabilities man is integrated in the process and is frequently subjected to unbalanced stress. Examples are the visual checking on mass production, acoustic tests in the ceramics industry and feeding functions in mechanical production.

In the handling of workpieces and tools, particularly in assembly operations, the human operator not only uses his motor abilities but also simultaneously his tactile and visual sensory organs for recording information.

The shortening of the non-productive time, during which most handling processes are carried out, was not considered an important problem to be tackled until a few years ago. The automation of handling processes in the production and assembly of components becomes more difficult the more frequently the handling operations change at a particular workplace. Here the use of industrial robots is a possible solution if clamping, orientating, checking and similar functions can also be automated or transferred.

The use of industrial robots for tool handling in spot welding, arc welding, coating or trimming also presupposes automation of the peripheral equipment (e.g. automatic sorting, automatic welding wire or coating material feed, automatic clamping, testing, etc.).

Thus, in the case of workpieces or tools which are not accurately positioned, or in the case of large workpiece tolerances, the use of industrial robots is not successful because suitable sensors for the detection of stochastic errors are not yet available.

257

6.1.1 Types and tasks of sensors for industrial robots
6.1.1.1 Definition
By sensors are meant those parts of a measuring device which are exposed directly to the phenomenon to be measured or detected. This definition is considerably expanded in some of the relevant literature. Sensors are used to detect stochastic influences in the area surrounding the industrial robot for measuring physical values and for pattern and position detection (6.1.).

Technical sensors process the signals of one or more detectors until complete information is obtained (e.g. orientation and position) for an industrial robot, information which can contain both simple yes-no decisions or analogue or digital data.

A sensor therefore consists of the detector and the associated signal evaluation, but signal conversion and amplification may also be carried out in the sensor (Figure 6.1).

Information detection by means of sensors may be tactile or non-tactile. Non-tactile sensors, for example, operate on the following physical principles: optical, capacitive, electromagnetic, inductive or fluidic.

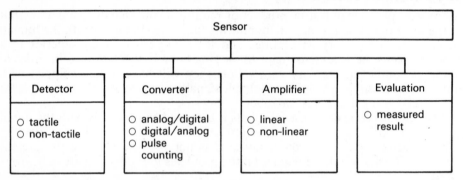

Figure 6.1   Sub systems of a sensor.

6.1.1.2 Types of sensors
Tactile and non-tactile sensors are mainly used for the different handling and machining operations with different workpieces and tools. The auditive sensors also listed in Figure 6.2 have so far only been used in production engineering for quality testing and for adjustment.

Visual sensors are particularly important for automation, which is why research work has been carried out on them for many years. Despite the importance which they have for automation, visual sensors have so far hardly been used in industrial practice, a situation which can be explained by the high cost, in some cases, and the low degree of reliability of the sensors. The more recent sensors are of more compact design due to the spectacular developments in electronic components, they are more reliable and have a longer life, and there is the prospect of further cost reductions. Figure 6.3 provides a survey of the main characteristics of the most important vision sensors.

Tactile sensors detect forces and moments or shapes by contact. Force or moment distributions, e.g. for machining processes or for assembly, are also detected. Some typical design forms of the many possible types of tactile sensors are shown in Figure 6.4.

Auditive sensors are microphones with which acoustic information is converted

258

Sensors			Evaluation	
	Design	Application	Principle	Boundary conditions
visual	semi-conductor sensors – point – linear – matrix vidicon tubes	position recognition condition testing classification	detection and processing of individual physical values probes	signal patterns may appear in multiple array or as a function of time
tactile	probe (pneumatic, electric, etc.) resistance strain gauge load cells (piezo, cap.) pressure sensitive plastics pin matrix	joining operations assembly tool monitoring position detection	detection and processing of n-dimensional patterns	large information quantities with visual sensors (television picture: 500,000 points, 30 million bits) data reduction – picture sections – object arrangment in preferred positions – optical aids reduction of processing time
auditive	microphone acceleration detector	quality control adjustment		– multiprocessors – special hardware

Figure 6.2   Survey of sensor technology (6.2).

to signal patterns. Sound in engines or ceramic products for example, may be recorded as a quality characteristic if a quality-specific sound pattern exists.

6.1.1.3   Task areas for tactile sensors

It is clear from workplace investigations that the applications of industrial robots can be extended if suitable sensors are developed to perform the sorting and inspection functions. Figure 6.5 shows the tasks which can be performed by an industrial robot with tactile sensors.

The assembly of workpieces with close tolerances (e.g. bolts or screws in appropriate holes) is often impossible because of the positioning uncertainty of industrial robots. Tactile sensors must detect the deviations of the parts to be joined

	Sensor	Size of light-sensitive area		Number of picture elements	Size of picture elements	Read-out frequency of picture elements	Useful spectral range	Price (1976) camera
		$\varnothing$	mm		μm^2	MHz	nm	DM
Tube base	Plumbikon	25	46	40. 000	50x50	max. 7	300- 800	1200-2500
Tube base	Multidiode vidikon	16	25	100. 000– 1. 000. 000	12x12	max. 5	400-1000	1200-2400
Semiconductor base	Photodiode matrix	3, 2	5	50 x 50	100x100	0, 5	200-1 100	15 000
Semiconductor base	CCPD-Matrix	6 x 6		100 x 100	60x60	10	200-1 100	ca. 15 000
Semiconductor base	CCD-Matrix	3 x 4		100 X 100	30x40	max. 5	400-1000	2 500
Semiconductor base	CCD-Matrix	7, 3 x 9, 75		512 x 320	30x40	3	400-1000	8 500
Semiconductor base	CID-Matrix	3 x 6		128 x 128	45x45	5	400-1 100	2 000

Figure 6.3   Survey of vision sensors (6.2).

259

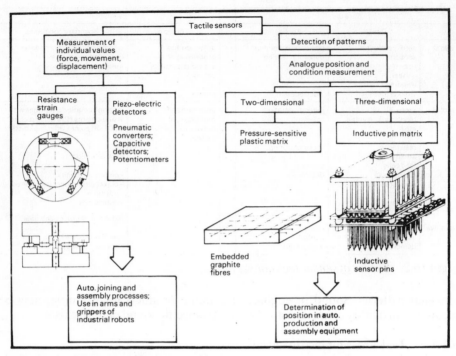

Figure 6.4   Different types of tactile sensors (6.2).

and enable the position and orientation of one of the parts to be joined to be corrected.

When assembling complete sub-assemblies, test processes must also be carried out and forces or moments limited to prevent damage, for example, to the components.

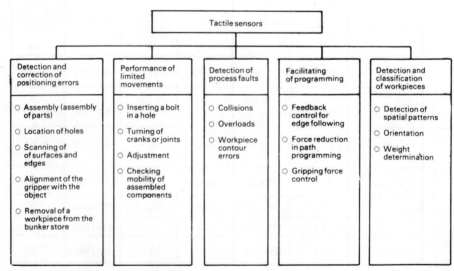

Figure 6.5   Possible applications for industrial robots with tactile sensors (IPA).

When automating assembly with robots in small and medium series production, the inspection functions normally carried out visually or by touch by human operators are performed by sensors. In the simpler cases, sensors can monitor the programmed movement cycle and interrupt the process in the event of faults, but they must control the assembly direct, e.g. by search movements or by position or pattern detection.

When operating with driven and non-driven tools which are guided by hand, contours must be detected visually and by touch. When trimming or settling, for example, some operations must be carried out by means of rotary grinding tools, which is physically very strenuous.

These operations must also be carried out by the human operator under difficult conditions (noise, dust, etc.), even in series production, since he has the visual control and sense of touch to detect both minor and major shape deviations and to machine them. When these operations are automated with a robot, the guidance of the tool must be supported by sensors which prepare information on forces acting on the workpiece so that the control system of the industrial robot can be suitably affected.

6.1.1.4 Areas of applications for non-contact sensors
The problem of orientating workpieces may be regarded as one of the main reasons why industrial robots are at present only being used in relatively small numbers, with the lack of flexibility of the devices in use so far being seen as one of the principal difficulties. At many workplaces visual tests are carried out as the method of quality control. In the case of automation with robots, these tests must also be carried out automatically. Further tasks for nontactile sensors include:
– detection of workpieces on suspension gear, etc.
– detection and monitoring of welded work (location of the start)
– measurement of workpiece dimensions and positions.

For small and medium batch sizes, workpieces are frequently conveyed between

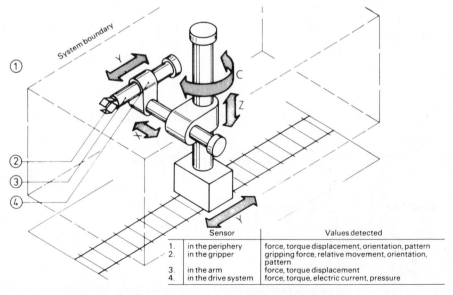

Sensor		Values detected
1.	in the periphery	force, torque displacement, orientation, pattern
2.	in the gripper	gripping force, relative movement, orientation, pattern
3.	in the arm	force, torque displacement
4.	in the drive system	force, torque, electric current, pressure

Figure 6.6   Arrangement of sensors on the industrial robot (IPA)

production equipment in bins, from where they cannot be removed directly by the robot. To do this, the workpieces must first be brought into a position and orientation known to the robot, i.e. defined. The purely mechanical orientation devices, which are not used for small batch sizes, for economic reasons, may be replaced by optical sensors.

In addition to the detection of the immediate surroundings of an industrial robot, e.g. with optical sensors for monitoring the working process, workpiece recognition and classification are the main tasks for non-tactile sensors. The sensors must enable a robot to be positioned, for example, according to optical patterns or enable it to pick from a bin, so that an industrial robot can also be used where the human operator was absolutely necessary because of his visual capabilities.

6.1.2 State of development

Sensors have been developed at several research institutes for various tasks, particularly in the USA, Japan and the German Federal Republic. So far, the possibilities shown in Figure 6.6 have been used for the arrangement of sensors on industrial robots, according to the tasks to be performed.

6.1.2.1 Tactile sensors

The automation of work processes which could previously only be carried out with the human sense of touch, for example in mating operations in assembly work, or in the handling of easily deformable workpieces, requires complicated tactile sensors with force and torque measuring devices. The need for developing tactile sensors was recognised some years ago, and there are some proposals for possible solutions.

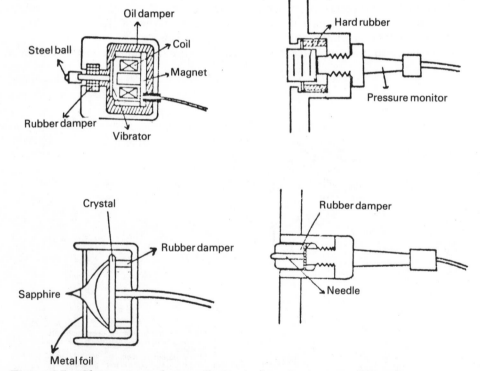

Figure 6.7   Chute sensors for installation in the gripping area (Waseda University).

1. Adaptive gripping force control

The first developments that had integrated tactile sensors in the gripping areas became known in 1972 at the Nagoya University in Japan. With these adaptive gripping force control systems were constructed which enable workpieces to be gripped with only a minimum force in order to be able to handle even delicate or easily deformable parts reliably and safely. Some of the sensors used and installed in the gripping area are shown in Figure 6.7 (6.3).

The Automation Research Laboratory of Kyoto University in Japan has developed a three-jaw gripper for upright cylindrical parts, each of which has a proximity sensor at the tip from which an air jet is emitted. The gripper is moved downwards for gripping and stops at a short distance above an object if at least one of the sensors reports 'contact'. The axis of the gripper then moves towards the axis of the cylindrical object, the direction required being determined from the sensor signals. At the same time the jaws open in steps until none of the air jets contact the object. Thus the position and the required width are obtained and the part can be gripped (6.4). Maschinenfabrik Augsburg Nürnberg have developed an adaptive gripping system which uses the weighing principle as its basis. Potentiometers are installed as measured value detectors in one of the two gripping jaws for recording the weight of the object and the gripping force exerted. Both values are fed to a controller which after comparison supplies the set value for an electric stepping motor. To avoid excessive gripping forces a slip sensor is also provided on one of the jaws. This sensor is in mechanical or magnetic contact with the object, and enables the coefficients of friction to be determined (6.5).

The use of such gripping sensors is mainly of interest on manipulators for research purposes and in nuclear engineering. For industrial robots a gripping force control system is sufficient in many cases, i.e. the much more expensive control system can be dispensed with.

2. Automatic assembly devices with sensors

An initial solution to the problem of automating simple assembly processes was proposed as early as 1972 by the firm Hitachi Ltd. (Japan), and has been improved upon ever since. The unit developed for automatic assembly is known under the name of HI-T-Hand (Figure 6.8).

The HI-T-Hand is capable of inserting a bolt in a suitable hole, even if the two axes do not coincide. By the use of a spring-loaded articulated unit, the hole can be detected with the workpiece (bolt) by means of different search patterns. The search time is approximately three seconds, which is shorter than the time taken by an unskilled worker. Bolts with a tolerance of 20 μm, and a diameter of ϕ 20mm, can

Figure 6.8
Bolt fitting system
HI-T-Hand (Hitachi) Ltd.

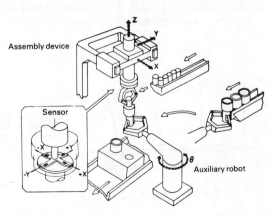

263

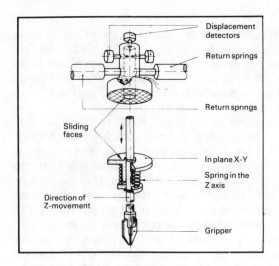

Figure 6.9
Tactile sensor on the
SIGMA assembly robot
(Photo: Olivetti)

Displacement detectors
Return springs
Return springs
Sliding faces
In plane X-Y
Spring in the Z axis
Direction of Z-movement
Gripper

be fitted. Such an assembly is limited to simple problems, i.e. two or three fixed bolts cannot be fitted into suitable holes simultaneously. The frequency with which such 'simple' problems arise, however, justifies this limitation [6.6].

A system has also been developed without drive [6.7] for fitting bolts into corresponding holes which performs the assembly process with vibration assistance where there are minor deviations in hole and bolt centres and large chamfers, making use of a low friction parallel guidance system.

A computer-controlled industrial robot assembly system with tactile sensors has been developed by Olivetti & Co. (Italy). The heart of this system is the SIGMA assembly robot developed by Olivetti, whose essential characteristics consist of its two independently operating arms, each with three degrees of freedom, and its equipment with force sensors for testing the assembly processes (Figure 6.9). A user-orientated language was developed specially for programming the computer, and the assembly robot has so far been used at about 40 different workplaces for assembling printed circuit boards [6.8].

The gripper joint is constructed so that small deflections are possible between the gripper and arm in all three directions – X, Y and Z – causing the production of elastic restoring forces. The deflection is measured by three displacement detectors, and this value is also a measure of the forces exerted by the workpiece on the component to be assembled. It is possible to determine forces of the order of 0.5 N with this. In most cases this sensor system is not used, however, for determining forces, but positions and deviations from them.

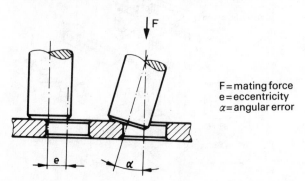

F = mating force
e = eccentricity
α = angular error

Figure 6.10
Angular error
produced when
applying an
inserting force.

264

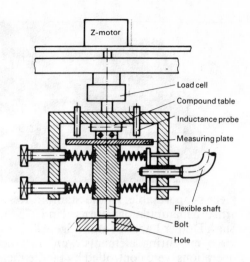

Figure 6.11
Principle of the
assembly device (IPA).

Z-motor

Load cell

Compound table

Inductance probe

Measuring plate

Flexible shaft

Bolt

Hole

An experimental device has been developed and constructed at the Institute of Production engineering and Automation (IPA) in Stuttgart, which automatically carries out an assembly operation (Figure 6.10) using the angular error produced by an insertion force. An insertion operation with cylindrical bolts can be controlled easily without expensive strategies or search aids. When a bolt is forced onto the corresponding hole, a tilting angle is produced which is a function of the inserting force and the eccentricity. An assembly device must be capable of detecting this angle and the insertion force, and of carrying out a corrective movement. The difference in diameters is decisive in calculating the possible eccentricity, as also are the radii which are produced at the point of contact, i.e. by chamfering the permissible eccentricity is considerably increased. With a diameter of 20 mm and a tolerance of 20 μm, the eccentricity is 0.5 mm if there is no chamfer; given the same tolerance and a chamfer of 0.5 mm on each component, an eccentricity of 4.5 mm is permissible [6.1].

The mechanism of the system shown in Figure 6.11 has very low weights to achieve high speeds and therefore short cycle times. The corrective movement is produced by stepping motors which are controlled by simple electronics without computer processes. This system is capable of fitting bolts with a tolerance of only 30 μm with a diameter of 20 mm, after less than 0.5 s "search time", without pre-determined search strategies having to be used, as in other systems.

During the insertion of a bolt in a hole where the axes are not exactly in line with each other, jamming can occur, with the result that even the smallest angles must be recorded and processed.

The assembly device is designed for fitting to an industrial robot, and therefore permits assembly work in which the industrial robot carries out coarse positioning. The actual insertion operation is performed by the auxiliary assembly device automatically, whilst the robot is stationary and continues its movement in the programmed cycle after the successful insertion operation (Figure 6.12). During the movement of the robot, in order, for example, to collect a new workpiece from a magazine, the assembly device can take up a defined central position from which the next search process is started.

3. Sensors for force-dependent control
In addition to the sensors in the gripper and for insertion, sensors have also been developed which measure reaction forces in the arm of the robot or in the work-

Figure 6.12
Auxiliary assembly
device fitted to an
industrial robot
(IPA).

piece support table. At the Stanford Research Institute (USA) force sensors in the arm of a manipulator were used in 1973 for force-dependent control. At the Charles Stark Draper Lab. in Cambridge, USA, the opposite route was taken initially, and force measuring elements were installed in a table. On this table insertion operations were controlled by tactile means using experimental workpieces, with the industrial robot (Unimate 2000) operating with a computer control system according to signals from the measuring table [6.9]. The advantage of the robot-independent force measuring device lies in the fact that no coordinate transformation has to be carried out. However, the measurement of the forces in the work table did not prove practicable since there were very many disturbing forces affecting the results.

When using a force measuring element in the arm of the handling unit the disturbing influences are much less marked, so that it is hoped to achieve better results here.

At the IPA in Stuttgart a sensor system has been developed which permits the use of industrial robots for trimming or the cleaning of castings. For trimming with rotary grinding tools, a 6-component force and torque detector was integrated in the arm of a handling unit whose signals are converted by a process computer to desired position values for controlling the handling unit. For this purpose detectors adapted to a handling unit were used which with high resolution allow adequate separation of the forces and torques produced. The handling unit controlled according to the sensor signals is able to follow a workpiece contour which is not precisely known (edge or surface) and in doing so can exert a predetermined force in a programmed direction [6.1].

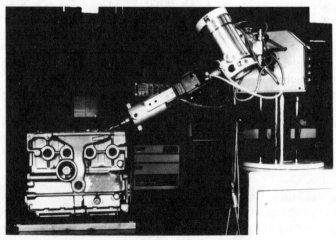

Figure 6.13
Industrial robot with
sensors integrated
in the arm for
trimming (IPA).

266

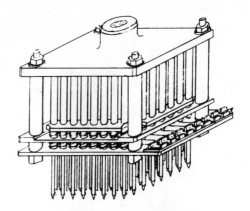

Figure 6.14
Tactile sensor
for pattern
recognition
(Nottingham University).

A programmable trimming system, with a handling unit for guiding a rotary tool, was designed on the basis of the requirements derived from workplace analyses, with special consideration given to the possibilities of controlling a handling unit. The design chosen, with a process computer for processing the signals from a force and torque detector, requires a program whose structure is determined by coordinate transformations. The relationship between the measured forces and torques is represented via the direction cosine matrix, which serves as a basis for the program. The program flow was simulated with the process computer used for preparing the program, and then tried in practical trimming tests (Figure 6.13).

The fettling of large castings or forgings has so far been impracticable because of the large tolerances, except with the aid of the human sense of touch. The use of handling equipment is only possible using a tactile sensor which permits force-dependent control for machining edges or surfaces with a defined force. The design using a computer for processing the detector signals provides a system which can in principle be used together with any handling unit which has the facility for external data feed. The sensor for fettling represents an important step in the direction of automating work on castings, as it has been shown that it is possible for a handling device with a tactile sensor to take over the sensory capabilities of man which are necessary for this operation.

In most cases patterns are detected by means of optical sensors, but tactile sensors with limited resolution may also be used for simple spatial patterns. At the University of Nottingham [6.10] a sensor has been developed which permits three-dimensional pattern recognition with a matrix of 8 × 8 sensory pins (Figure 6.14). Pattern recognition with tactile sensors has only a very limited application, and as far as resolution is concerned it is far surpassed by the non-tactile systems.

6.1.2.2 Non-contact sensors
Today industrial television cameras and diode arrays are mainly used for picture detection as sensors for arranging workpieces by position, orientation and pattern recognition. The other remaining non-tactile sensors are far inferior with regard to resolution. As early as the 1960's, the Stanford Research Institute (SRI), USA, began work on the development of research robots controlled by television cameras. The problem of pattern recognition was tackled with the experience gained, and technical solutions presented already a few years ago.

The work by the same Institute in the field of pattern recognition with grey level image processing was begun only a few years ago because the cost of computing time and computer capacity was enormously high and there were no prospects of reducing this cost. The knowledge of this led to a situation where almost all present

267

Figure 6.15
Optical correlator
for industrial robot (IPA).

research work is concerned only with "binary" black-and-white pictures. For applications in space and marine technology, processes and methods for the analysis of three-dimensional pictures have been developed, but these are very involved, computer intensive and slow, and cannot therefore be used as sensor systems on handling equipment.

Not until later was work begun in the Federal Republic on the development of sensors, with a clearer view of the problem, and an attempt was made in the recording of information to reduce the data quantity. Since most conventional industrial robots still operate without computers, the sensors were developed as independent systems.

1. Sensors for recognising moving workpieces

This procedure is also used in the recognition device, a so-called optical correlator, which was developed at the Institute of Information Processing in Engineering and Biology (IITB), Karlsruhe. The unit is suitable for recognising circular patterns such as those obtained from round workpieces or holes, and for measuring the position of the centre of the circle. This unit, together with a conveyor belt as the orientation device for an industrial robot, has been used at the Institute of Production Engineering and Automation (IPA). Round, disc-shaped workpieces of varying diameter are fed by a conveyor belt past the optical of the the correlator (Figure 6.15).

The "internal model" consists here of a fast rotating exchangeable disc on which are arranged transparent photographs of the objects to be detected, around the periphery. If a workpiece which corresponds to the photograph enters the "catchment area" of the unit, both the photograph and workpiece are covered at a defined point of time. This is recorded by a photodiode, which initiates simple

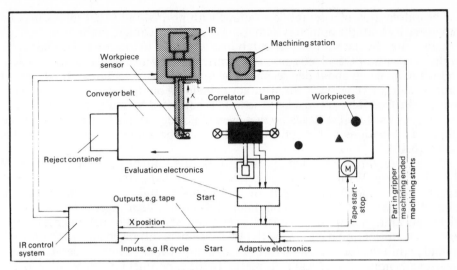

Figure 6.16. Basic structure of an industrial robot with correlator (IPA)

electronics to determine the position of the centre of the circle and transfer this value to the control system of the industrial robot (Figure 6.16).

The movement of the industrial robot is only initiated when a "correct" workpiece, whose diameter corresponds to that of the picture, passes the correlator. Parts which are too large, too small or even non-circular, are detected as "incorrect" and are left to pass on the conveyor belt. The same principle can also be applied to the position measurement of holes in workpieces. The position of a workpiece can be determined by means of two or three specified holes in any shaped workpiece, and the gripper of the industrial robot controlled accordingly [6.11].

Helsinki University has developed a recognition system in which several work-pieces can be present in the detection range of a diode array [6.10]. A very simple array with only 50 × 50 elements was used so as to achieve a simple system of data processing. The resolution of the system, which was tested with an ASEA industrial robot, is very limited, but in addition to stationary workpieces, moving workpieces can also be detected and gripped (Figure 6.17).

2. Sensors for detecting stationary workpieces

Optical scanning for three-dimensional objects has been developed at Edinburgh University on the basis of the artificial intelligence methods, in which the shape of the object is detected without information. By correlating two stereoscopic TV contour images, a shape describing network is placed iteratively over the object (multicontour process) [6.12]. Although the method is not yet fully developed, it is nevertheless an attempt to find methods, from among the much too involved methods in the field of artificial intelligence, whose cost is justifiable in industrial production.

The University of Rhode Island (USA) has developed an optical workpiece detection system with which also overlapping workpieces can be detected [6.10]. For this purpsoe a TV camera is used whose binary images allow a statement to be made on the position and orientation of the workpieces on the basis of line-by-line evaluation (Figure 6.18).

The automation of assembly operations with normal industrial robots may be supported by a simple optical system using a small TV camera in the gripper. This system is used by the Stanford Research Institute (SRI) to locate a hole. The image evaluation with a PDP 11/40 computer is carried out by means of a "binary" black-and-white image in about one second. The search process requires a total time of approximately five seconds, the accuracy depending on the position control of the industrial robot. The approach to the hole is corrected by multiple evaluation of the image, in successive stages, so that even major deviations between the desired and actual positions of the industrial robots can be compensated by using this system [6.10].

The only requirement is that the hole to be located still be in the field of vision of the TV camera. With a similar objective, work is being done at the Mechanical Engineering Laboratory (Japan) on an experimental system which is also controlled with a PDP 11/40 computer. To achieve high degrees of accuracy a highly accurate manipulator is used, which can be moved in increments of 0.01 mm. The information recording with two TV cameras arranged next to the gripper should permit the use of this system even for complicated assembly operations with rotationally non-symmetrical workpieces. In the experiment it was possible to fit a rectangular workpiece into a corresponding recess with a tolerance of 0.1 mm, the workpiece being painted white to improve the contrast. The mating and locating process, requiring a total time of approx. five minutes, is much too slow for practical application.

For the orientation of workpieces, a television system was developed at the IITB, in which the parts are moved horizontally on a conveyor belt against a simple stop, and there assume a series of discrete positions (Figure 6.19). A television camera is used as the image recording unit, and from the image produced, only a few lines are evaluated in a microprocessor for orientation and position determination. Only a 1.5 k memory is required for the program and data due to this data reduction in the image scanning. The system is programmed by "showing" the workpieces in their possible positions. For workpieces which cannot be guided against a stop in a

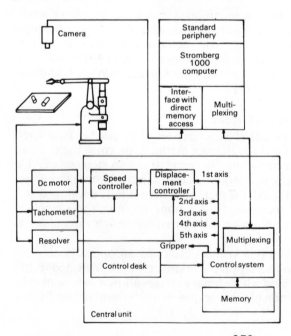

Figure 6.17
Diagram of system with
ASEA industrial robots
(Helsinki University).

*Figure 6.18
Actual and "binary"
image of overlapping
workpieces
(University of
Rhode Island).*

defined position, a similar system has been developed at the IITB which is able to recognise a much broader workpiece spectrum [6.13].

The detection process consists of two stages. Firstly, the surface area and area centre of gravity of the plan image of the workpiece area coincides with that of the programmed workpiece, concentric circles are placed around the centre of gravity. The position of the points of intersection of these circles with the workpiece contour is a further recognition criteria, and for recognised workpieces it is a measure of the angle through which the workpiece is turned in relation to a defined reference position. The sensor is programmed by previously placing the workpiece in all possible positions.

Figure 6.19.   Television sensor system with microprocessor.

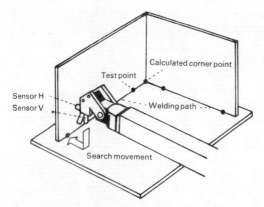

Figure 6.20
Sensor for detecting
welding paths
(Photo: Komatsu).

For welding, a sensor has been developed in the Hitachi Research Laboratory which enables welding work to be carried out even where the workpieces are not accurately positioned. The sensor shown in Figure 6.20 consists of two inductive proximity switches, sensor H and sensor V, whose axes are perpendicular to each other and both of which form an angle of 45° to the welding electrode. Before the welding process, the industrial robot scans the welding seam with the sensor, the position of the seam only being roughly known initially, at several points. If the seam is a straight line, two points are measured from which the industrial robot determines the welding path. Curves can be approximated by measuring several points. Corners are obtained from the point of intersection of two straight lines and are also welded automatically. The system is being used at Komatsu for welding excavator parts [6.4].

A device was developed by the Nagoya University, Japan, for guiding the electrode of a welding robot along a curved plate to achieve a greater degree of accuracy than with a mechanical sensor: beamed light pulses are thrown onto a mirror with surfaces arranged in a triangle and rotating synchronously with the frequency of the source, which momentarily illuminate a point on the surface of the plate. Reflected light is detected by a receiver at a certain distance underneath the mirror, for instance, in a defined direction. The distance is determined by time of flight measurement [6.10].

A workpiece recognition system which is equipped with a special optical device has been developed by Olivetti (Figure 6.21). The position and orientation is

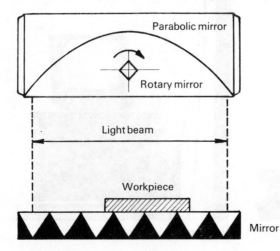

Figure 6.21
Optical recognition
device
(Photo: Olivetti).

272

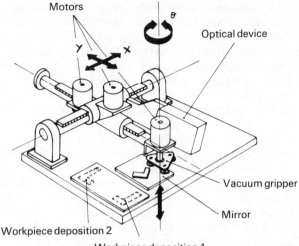

Figure 6.22
Workpiece recognition
system (Photo: Olivetti).

Motors

θ

Optical device

Y X

Vacuum gripper

Mirror

Workpiece deposition 2

Workpiece deposition 1

determined in one axis by means of a beamed light which is reflected onto a photo-electric cell. In the second axis the complete device is moved by a stepping motor, so that the working surface, measuring approx. 200 × 200 mm, is scanned in 0.5 s (Figure 6.22).

The recognition system is particularly suitable for flat workpieces, and combines low-cost construction (low memory space requirement) with a high degree of flexibility [6.10]. The unit, which is designed for an industrial environment, is equipped with three degrees of freedom, and is simply programmed by "showing" a workpiece.

3. Sensors for synchronisation with a continuous conveyor device
The GTE Laboratories (GTE: General Telephone and Electronics Corporation USA) have developed an industrial robot system controlled and monitored by a

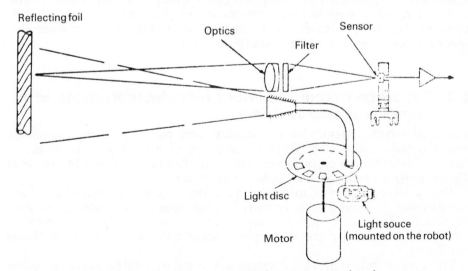

Reflecting foil

Optics

Filter

Sensor

Light disc

Light souce
(mounted on the robot)

Motor

Figure 6.23. *Sensor principle for continuously rotating overhead conveyor.*
(Works photo: GTE)

PDP-11 mini-computer, with sensors for guiding an industrial robot on a continuously running overhead conveyor, which has already been in opeation for two years [6.14]. The industrial robot (AMF-Versatran) is used to pick up television tubes from a cycled belt, rotate them through 180° and deposit them on a continuous overhead conveyor. The transfer from the industrial robot to the overhead conveyor requires an accurate determination of the position of the workpiece carriers of the overhead conveyor. This is achieved optically by means of a light source installed next to the overhead conveyor, which emits light impulses which, on encountering a workpiece carrier, are reflected by reflecting adhesive tape and received by a photodetector (Figure 6.23). The robot is designed so that it can be moved to achieve synchronisation between the movements of the overhead conveyor and the robot. The mini-computer calculates the exact location and speed of transfer of the tubes by means of the signals from the photodetector, and transmits appropriate commands to the robot. In addition to this function, the computer is also used to control the entire system. The mini-computer is programmed by means of the ALFA interactive language specially developed for such automation operations by GTE Laboratories.

This system processes 1000 television tubes in 8 hours without major faults or breakdowns, and is therefore used very economically.

The development of sensors for industrial robots has been progressing on a worldwide basis with varying intensity for several years. However, the first results were too expensive to be used economically in production. The new developments are concentrated more closely on the relevant problems in the tasks they are designed to perform, with the result that solutions promising a measure of economic success are already being tested in prototype applications.

The development of simple tolerance compensation systems operating without a computer will also allow the application of simple industrial robots to be extended to more complex tasks which even in the long term cannot be completely replaced by computer-controlled industrial robots.

The newly developed industrial robots are frequently provided with computer control systems so that no additional computers have to be installed for the use of sensors which require extensive signal processing. Instead, the entire system can be controlled by one computer. Improvements and simplifications are basically still necessary for most of the sensor systems so far developed before successful application can be expected in industrial production.

6.2 Experimental constructions for industrial robots with complex sensor systems

Industrial robots currently on the market carry out very simple, frequently repetitive tasks in most cases. Only the use of optical or tactile sensors of the type described permit better use of the flexibility of freely programmable handling units already provided, giving a wider application for industrial robots.

Thus a large number of assembly operations require coordinated eye-hand movements. Accordingly an industrial robot requires in such cases good coordination between the signals of optical and tactile sensors. This coordination can now be provided by means of computer control systems, but these are still being developed.

These recent industrial robot developments are being used for handling, testing and assembly operations, particularly in small-batch production, since they no longer require a determined or structured [6.15] periphery. Within the foreseeable

Figure 6.24   SIRCH industrial robot (Photo: University of Nottingham).

future they will be able to economically handle workpieces which are orientated in a fairly random fashion. These possibilities will then also be sued for more complicated assembly operations, and it is expected that these newly developed systems will be ready for the market, either individually or working together in groups, within the next 5 years [6.16]. The extent of this will depend largely on their reliability, their economic application possibilities and on the capital available in the industrial companies.

But despite these long-term prospects, there is still a great deal of research and development work to be done. Some typical developments and experimental constructions are described in the following, primarily observed in the past few years in the USA and Japan. In Japan there are as many as 70 government laboratories or laboratories attached to universities which are working directly in the field of industrial robots or in related fields [6.17, 6.18]. From the very beginning, German development work was more industrially orientated, in order to arrive as quickly as possible at economically practicable solutions. However, it is already being recognised that – as was the case with the first generation equipment –developments of industrial robots with complex sensor systems and suitable information processing, aimed at economic industrial application in the near future, are increasingly being tackled also in Germany.

6.2.1   Testing of sensor systems
The developments at Stanford University, USA, may be regarded as the early pioneer work for future industrial robots. The first attempts at almost complete recognition also of coloured patterns were made using a TV camera which operated with interchangeable lenses in the revolver and colour filters for detecting four-coloured building blocks. Edges and corners of the building blocks in the visual field were highlighted by means of a computer program which was naturally involved and operated very slowly, and their position in front of the industrial

robot determined. Following predetermined rules the building blocks could then be assembled to form a tower with a predetermined colour combination [6.20].

By contrast to these laboratory experiments, which were initially far from leading to an industrial application, a further development was presented in 1973, with which a water pump consisting of two parts, a seal and screws, could be assembled [6.16]. The force was measured indirectly in this case by means of the recorded motor currents. Among other things it was used to fit the screws into the tapped holes. If the centre-lines of the hole and of the screw did not coincide sufficiently accurately, resistance in the form of forces was recorded when the screw was lowered. The screws could then be inserted in the threaded holes by means of a search operation.

The further development in the field of pattern recognition using optical sensors went a different and simpler way, however, which led more quickly to industrially applicable problem solutions using a binary black and white image processing method.

An industrial robot was developed at Nottingham University, England, for assembly operations, under the name of SIRCH [6.21], which was able to recognise flat workpieces of any shape (Figure 6.24). For this purpose the industrial robot is provided with a gripper turret which carries three grippers and is additionally equipped with a rotating TV camera. The camera is used for recognition and position determination of different objects which are dispersed randomly on an illuminated area. Centres of gravity, areas, circumferences and other dimensions of two-dimensional images are determined for the purpose of recognition and position determination, and compared internally with patterns which were previously stored by simply "showing" the workpiece in its different possible positions on the table. The industrial robot selects the gripper suitable for handling the workpiece and brings the part to a predetermined location, in a position

Figure 6.25   Experimental set up for bin-picking (Photograph: University of Rhode Island.)

Figure 6.26
Layout of a workplace
for automatic carbon
brush assembly
(Photo: Mitsubishi
Electric Corp.).

required for assembly. The video signals are processed and controlled by means of a digital computer with a memory capacity of 8 k.

The experimental set up developed by the University of Rhode Island, used in 1978 for bin-picking, is shown in Figure 6.25 [6.22]. The optical sensors described in section 6.1.2.2 were used for this purpose. In the experimental set up, a 6-axis industrial robot was used. Simpler elements from machine tool construction were used for its 3 linear axes, and a further 3 rotational manual axes were fitted to its projecting Y axis. A diode array camera is located on the cantilever arm between two lamps, with the camera "looking into" the bin. A suction gripper, whose hose is guided in a helical spring, picks up the workpiece, so that any workpiece positions are permitted in the box for gripping. This gripping process operates first time in 60 to 80 out of 100 cases. To establish the position in which the workpiece was gripped, it is shown to a further fixed camera (arranged on the right in the photograph). If the workpiece has to be re-gripped, it is deposited on a support also provided with a vacuum gripper. In order to feed the workpiece into a machine (not included in the experimental set up), an additional gripper can be flanged onto the robot, to provide a simpler feeding method into a machine than is possible with the present robot design. To establish the accuracy of the picking up of a workpiece in the experimental set up, the workpiece position in the gripper is measured in a measuring device (in the foreground of the photograph) with 6 displacement recorders.

The picking up, recognition and positioning of a workpiece, still takes about 30 s in this experimental set up, which must not be confused with a pilot application for industrial use: rather this is basic research for which the equipment developed only represents an experimental unit which is certainly capable of further optimisation.

As early as 1973 a handling unit controlled by a mini-computer was developed by the Mitsubishi Electric Corp., Japan. This system is equipped with a television camera and is capable of locating objects within straight-line limits, picking up certain objects and depositing them on a moving conveyor device. A television camera above the industrial robot determines the number of objects and their approximate positions. After an evaluation of this video picture the arm is controlled so that a second television camera in the gripper is brought over the object and gripping can be effected under controlled conditions by an evaluation of the second video picture.

A more advanced system had been tested since 1975 for the assembly of carbon brushes in d.c. motors in the laboratory [6.23]. In addition to the industrial robot, a mini-computer, with an 8k memory PDP-8/1), two television cameras and a simple assembly machine form the essential modules of this system (Figure 6.26).

Figure 6.27
Experimental setup
for the assembly of
electric generators
(Photo:
Charles Stark Draper Lab.)

One of the television cameras is secured to the industrial robot arm and is therefore also moved. It is used for location and position recognition of the carbon brushes which are ready suspended on a rod. The other camera is integrated in the gripper and is used for fine recognition purposes. The robot, which has seven degrees of freedom, and is driven by stepping motors, positions the brushes and the brush holder in the assembly machine and magazines the assembled part. The assembly machine then carries out the actual fitting process. The mini-computer is used to control the whole system. To enable systems such as these to be used in production, however, considerable development work is still required, for in this case automatic assembly takes 12 times as long as manual assembly.

The results of research gained over several years in the field of computer-controlled automatic assembly of small and/or changing batches were illustrated by the Charles Stark Draper Laboratory, USA, in an experimental setup for the assembly of electric generators (Figure 6.27) [6.24]. The assembly system consists of an industrial robot with four degrees of freedom and associated periphery totalling 12 orientating devices, with automatic changing of 6 of the tools and grippers required. The industrial robot assembles the 17 component parts of the generator by means of two different fixtures from one direction only. To avoid jamming or wedging of the bolts in holes, which may occur as a result of workpiece tolerances and positioning errors of the robot, the robot is provided with a compliant hand joint which, within limits, is able to compensate for tolerances or positioning errors without the use of sensors and auxiliary drives. Two minutes 42 seconds are required to assemble the electric generator, which was chosen as the object of investigation mainly because it can be assembled from one direction only.

The assembly time could be reduced to 1 minute 5 seconds by improving tools and fixtures and by simple structural product modifications. The assembly time can in this case be reduced even further by organisational measures such as cyclic assembly (see also [6.25]), with similar subassembly operations being carried out initially and in stages on several generators before a tool or gripper is changed. This would have the effect of distributing the non-productive times resulting from gripper changes evenly over several items. Without these measures, the tool change times amount to about 30% of the total assembly time for one generator. Similar

278

Figure 6.28
Flexible sorting device
with camera and
industrial robot.
(Photo: SRI
International.)

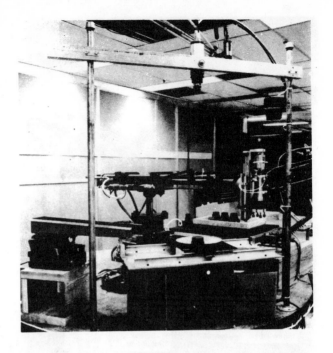

investigations by the IPA for reducing non-productive times for gripper and tool changes cna be reduced by about 90% as a result of such process organisational measures [6.25].

The economic considerations resulting from the testing of the experimental setup showed that programmable assembly systems with single-arm industrial robots should only be used for the assembly of smaller batches (less that 50,000 items per annum), if the single-arm robot can assemble each component part (as can the human operator) in about 3 to 5 seconds. In the case of larger annual batches, it is therefore recommended that several robots or multi-arm robots be used.

The research and development work of SRI International, USA, forms part of an on-going development programme under the name of "Machine Intelligence Research Applied to Industrial Automation", as also does the work done at the Charles Draper Lab. The different software and hardware modules were each integrated in different experimental setups to test them and demonstrate their operability.

As early as 1974 a pattern recognition system was proposed, which is able to distinguish workpieces according to their contour [6.26]. The system is able to establish the position of one or more workpieces and the completeness of a workpiece.

"Binary" images are used to recognise the parts, so that a PDP-10 mini-computer is able to process the data. The storage requirement can also be reduced by a suitable reduction in the workpiece characteristics evaluated. Instead of the television camera previously used, diode arrays are employed to produce what have so far been very good results.

For position recognition of parts suspended on a magnet the Unimate swivels over a diode array and deposits the part in different places, without any time delay, according to the recognised position.

The recognition process is programmed by "showing" the possible workpiece

Figure 6.29
Automatic assembly of
a compressor housing.
(Photo: SRI
International.)

positions. The picture used for the recognition is visualised on a screen for the programmer.

In 1978 SRI International presented another development of a flexible orientating device [6.27]. By using the programmable setup shown in Figure 6.28, three-dimensional workpieces which, because of their size, cannot be orientated in a vibratory bowl feeder, can be offered to an industrial robot in the correct (location and) position. The arrangement consists of a small vibrating conveyor, an optical recognition device, an industrial robot from the firm Auto-Place, and the actual orientating device. The latter contains three cylindrical chambers, over which a slide of circular section travels, and which is able to push the parts from one chamber to another. The bottom of the first chamber can be lowered to certain drop heights programmable for the workpieces. A workpiece drops into this chamber on to the bottom in the top position from the vibrating conveyor. The slide then brings the part to the back illuminated, rotary cover plate of the second chamber, on which the position and orientation of the workpiece are established by means of the optical recognition device arranged above the chamber. The latter consists of a diode array camera (128 × 128 elements), an LSI-11 micro-computer and a suitable inface-preprocessor. If the part is recognized as being undesirable, it is transferred from the slide into the third chamber. If the part is correct but in an incorrect position, the plate in the second chamber rotates the workpiece into a suitable position, from which the workpiece is able to drop onto the lowered bottom of the first chamber. For this purpose the part is pushed from the second chamber onto the edge of the first chamber until it drops. This alters the workpiece position so that after another recognition process on the rotary plate the part can be offered to the robot in the correct position. If necessary, this throwing-back process can be repeated several times if several stable positions of the workpiece are possible. This application is therefore a combination of optical sensors and utilisation of the workpiece behaviour.

280

In another experimental setup a compressor housing, consisting of compressor, cover and eight screws, is assembled. In the first version (Figure 6.29) this assembly device operates with a Unimate 2000 B, but in the absence of a simple unit for which the application was actually planned, the Unimate was limited to movements between 8 fixed points. The unit is therefore simulated with only 4 degrees of freedom. A gripper was fitted to the arm of the Unimate, and on the pneumatic cylinder of the gripper a linear potentiometer is attached for controlling the gripping force. The experimental setup also includes a programmable X-Y table, a diode array camera, peripheral mini-computers and an LSI-11 as the control computer for the Unimate. The periphery consists of screw magazine, screwdriver and two devices, one for holding the screwdriver, the other for receiving the cover. Another task was to demonstrate that for automatic assembly with the aid of optical sensors it is possible to manage with a minimum of devices and mechanisms in the periphery.

After a single calibration of the camera and compound table a housing is deposited in any position on the table. The camera records the image of the housing in plan view; the computer analyses this image, calculates the position coordinates and the orientation of the housing and causes the table to bring the housing into the centre of the visual field of the camera. The Unimate then grips the cover and places it onto the housing, whilst the table is released to compensate for position inaccuracies during assembly. The computer then analyses the camera image of the fitted cover, calculates the position of the first threaded hole and again lets the table move this to the centre of the field of vision, the release being first cancelled again. The robot then grips the screwdriver with the gripper, picks up a screw and places it in the threaded hole. The screwing operation itself takes place again with the compound table released. When the screwing operation is completed, the computer checks by means of the camera image whether the hole in the cover is closed, i.e. whether the assembly operation has been successfully completed. If this is not the case, the screwing operation is repeated. This process is carried out for all 8 holes, whereupon the screwdriver is again put down. The total assembly time is about 160 s. In other experimental setups the screwing operation is carried out by a robot of the Auto-Place 50 type, with an additional rotational axis and a screwdriver secured to the arm. The Unimate here removes the housing from a moving conveyor belt, and the covers from a container, by the use of optical sensors. In this container the covers do not lie in arbitrary positions, but they nevertheless do not form a defined arrangement.

The Watson Research Center belonging to IBM is also concerned with computer-controlled assembly, particularly for small parts [6.28]. For this purpose an hydraulically-driven industrial robot was constructed with Cartesian coordinates. A unit with four hydraulic cylinders, perpendicular to the direction of motion, moves the arm over a cam plate. The unit has three rotational gripper axes and a parallel gripper with tensile strain gauges for measuring forces and torques. This sensor arrangement enables difficult assembly operations to be carried out without the aid of optical sensors.

The signals processing and control are carried out by means of a computer of the IBM 7 type. Initial application experience was gained in the assembly of a subassembly consisting of 28 parts for an electric typewriter which was assembled in 6 minutes. The position of the fixture into which the component parts are placed is here recognised by tactile means.

6.2.2 Developments in and on multi-arm industrial robots
Developments for coordinated movements of multi-arm industrial robots are of

particular importance for assembly operations. The first two-arm industrial assembly robot available on the market, the SIGMA MTG, from Olivetti, is naturally also the subject of extensive investigation by research institutes. This work is being done in Italy at the Universities of Florence and Pisa, and at the Milan Technical University, and in Germany by IPA-Stuttgart.

In Florence primarily programming methods are being developed for optimising the movement processes for assembly. The object of this work is the maximum use of the time of the robot by the optimum distribution of the individual movement and assembly steps over the two arms of the industrial robot, by suitable determination of the assembly sequence and by minimising the waiting times of the arms which are required to prevent collisions of the two arms [6.29]. These problems are solved by means of the well-known methods of Operations Research or heuristic programming. The programs are written in FORTRAN IV. Owing to the large number of individual operations to be considered in assembly (in most cases about 100), a mini-computer, with a storage capacity of 16 k and a disc unit as the mass memory, requires about 10 mins. for one computer run.

At the University of Pisa the dynamic behaviour of both the software and the hardware of the SIGMA industrial robot is being investigated by means of digital simulation [6.29]. A SOFT program module incorporates the logical software, whilst in a HARD subprogram the kinematics and dynamics of the simulated system, i.e. the SIGMA industrial robot, are represented as the software model. The computer is able to reproduce all the position, speed and acceleration configurations of the real system in performing an assembly cycle by means of the HARD subprogram. The interaction between hardware and software is monitored and controlled by a supervisory program section, which also provides all the characteristic data on the simulated assembly operation, namely speeds, completion times, etc., and prevents collisions of the arms. Modifications of both the hardware and the software may be simulated and tested by means of this program package.

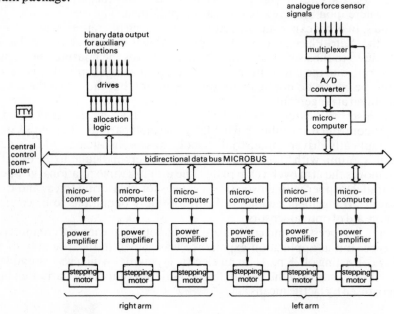

Figure 6.30   Control structure of the SUPER-SIGMA. (Photo: Politechnico di Milano.)

At the Institute for Electrical Engineering and Electronics in Milan, work is being done on a new control system under the name of SUPER-SIGMA for the SIGMA industrial robot on the basis of microprocessors. A 16-bit mini-computer is being used as the central control unit. Several micro-computers are arranged between this computer and the mechanical hardware of the industrial robot, and are connected to the central control unit by means of a special bidirectional data bus, which is termed a MICROBUS (Figure 6.30) [6.29].

To avoid positioning problems, stepping motors, with which, as is well known, no closed position control circuit can be constructed, must be controlled fairly accurately. For this reason each stepping motor is assigned a microprocessor whose task is to accelerate the motor after an optimised movement and then to retard it.

The force sensors contained in the SIGMA industrial robot also require monitoring because of their analog nature. Their signal output must be digitalised with an A/D converter, to enable it to be further processed in digital form. As only one such A/D converter is used, the sensors are interrogated by means of a multiplexer. A very simple micro-computer, which incorporates only one ROM as the memory, is also used for controlling the grippers, screwdrivers, etc., essentially because of the modularity of the control concept as a whole.

At the Fraunhof Institute for Production Engineering and Automation (IPA), Stuttgart, a programmable assembly system has been constructed, consisting of a SIGMA industrial robot as the handling unit (Figure 6.31). To enable this programmable assembly system also to be used for small batches, new types of peripheral devices are being developed. For rapid resetting of the assembly system standardised plates on air cushions, for example, may be moved into the workspace of the robot and there indexed. The product- and workpiece-specific grippers and tools are received by magazines at the sides of the workspace. The arms of the SIGMA robot were equipped with tactile gripper/sensor systems consisting of an interchangeable flange, which is able to pick up the grippers and tools from a magazine by means of a standardised flange on the gripper, and place them there again on completion of the assembly cycle. The actuating device for the grippers and tools is integrated in the interchangeable flange, so that when the grippers or tools are changed only simple-mechanical components are changed. The problem of transmitting electrical or pneumatic energy from the arm of the industrial robot via the "interchangeable flange" interface to the gripper has therefore been

Figure 6.31
Experimental structure
of a programmable
assembly system
with claw change and
replaceable working
plates. (Photo: IPA.)

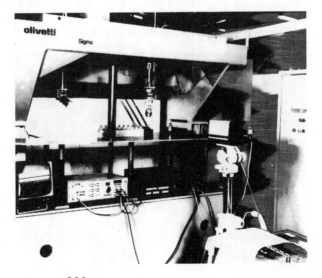

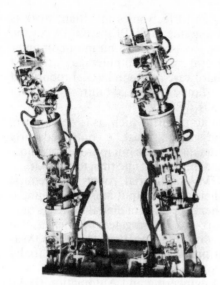

Figure 6.32
Two-arm "MELARM" industrial robot.
(Photo: Mechanical Engineering Lab.)

bypassed. The interchangeable flange is suspended on a kinematic setup which with a total of 6 degrees of freedom enables the robot arm to react sensitively to assembly forces. These forces can be measured by displacement detectors and evaluated by the computer of the robot. The compliance of the partially air-cushioned kinematic structure can be switched from the "soft" to "hard" condition for applying greater insertion forces. The entire tactile gripper/sensor system was of very light construction (12N) to keep the actual weight handled as heavy as possible.

In addition to the tactile control of the assembly processes, a television sensor was used by IPA to monitor the assembly process. The micro-computer only monitors a few preselectable lines of the television picture outside the industrial robot control system for this purpose. The video signal can therefore be processed very quickly and with very little computing involvement.

The government research institute, the Mechanical Engineering Laboratory (MEL), Japan, has developed a prototype for an industrial robot under the name of "Melarm" [6.17, 6.18, 6.30] as part of its research and development work in the field of mechanical movement devices, monitoring of moving objects with optical sensors and pattern recognition as well as tactile sensors. This industrial robot has

Figure 6.33
Experimental setup for
testing software strategies
for co-operatively working
arms on industrial robots.
(Photo: Electrotechnical Lab.)

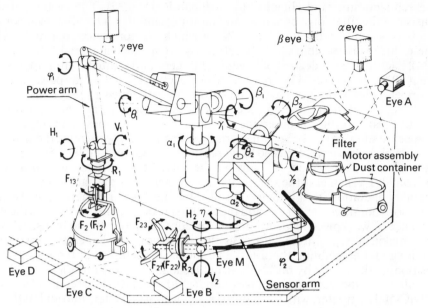

Figure 6.34   Experimental setup for assembling a vacuum cleaner. (Photo: Hitachi Ltd.)

two arms, each with 7 joints, and a gripper (Figure 6.32). A force sensor is installed in each joint; the grippers are equipped with several tactile sensors. Both arms perform coordinated movements by means of a mini-computer. In conjunction with a device for movement detection, which can be fitted to the arm of a human being, the industrial robot can also be controlled as a master-slave system. Further work on this project provides for the coordinate movement of the two arms in connection with a three-dimensional pattern recognition system. The cost of developing such complex systems is evident from the fact that this advanced Melarm system is not expected to be ready for use until 1985 [6.17]. Foundries, among other users, are then foreseen as a possible industrial application.

Investigators are also being conducted at the Electrotechnical Laboratory in Tokyo into the coordination of arms of multi-arm industrial robots [1.67, 6.18]. For this purpose, two independently operating robot arms, each with 7 degrees of freedom, are being used (Figure 6.33).

In addition to position and speed transmitters, a control system is provided for the torques for each axis. The movements are controlled by the fact that each drive unit applies a torque which is just large enough for errors which would otherwise occur between the actual and desired positions as a result of elastic deformation to be almost eliminated. The movements of the arms are in this case initiated by cable controls. The drives themselves operate by electric motor. Both arms are controlled by a CNC which is designed as a continuous path control system with linear interpolation.

A PDP-12 is used as the computer. Work such as sawing pieces of wood, joining wooden components and hammering nails can be carried out with this industrial robot.

One of the most interesting, and so far the most expensive, examples of the present state of development is represented by an experimental device for the automatic assembly of vacuum cleaners, which has been constructed in the

285

research laboratories of Hitachi Ltd. in Japan [6.31, 6.32]. The vacuum cleaner consists of the filter, dust container and motor assemblies. The filter itself consists of a plastic frame and a dust bag, both of which are connected to a rubber ring. These three subassemblies are assembled by a two-arm industrial robot (Figure 6.34). Each arm has 8 degrees of freedom, 3 fingers, and is fitted on its grippers with about 30 tactile sensors. The arm shown on the left in the figure, the power arm, has a gripper which is designed to lift the parts to be assembled and keep them in this position. A television camera is integrated in the gripper area of the right arm, the sensitive sensor arm, for observing the objects from all sides. The fingers of its gripper are suitable for recognising parts by scanning, and for picking them up. The vacuum cleaner parts are placed in no particular order on a table. The position of these parts is detected in 16.7 ms by the interaction of two vertical and one horizontal TV camera, the eye in the sensor arm and the tactile sensors. The assembly operation itself is then controlled by four further cameras, (one vertical, three horizontal, the sensor eye and the tactile sensors. Figure 6.35 shows the basic sequence of workpiece recognition and the control system of the industrial robot when picking up the filter. Figure 6.36 illustrates the control operation involved in handling and assembling the filter. The mode of operation of the system when positioning the motor by means of visual feedback is shown in Figure 6.37. The complete assembly operation takes about 120 seconds. It is controlled by an HIDIC-500 computer, with the industrial robot control system and an HIDIC-150 mini-computer as subordinate units monitoring the movement processes and the tactile information process. Figure 6.38 shows the operating principle of the control system. The image processing is carried out with both binary black-and-white pictures and with grey-key pictures in 256 brightness stages. The image memory therefore consists of a memory with 256×256 bits, and two $8 \times 256 \times 256$ bit memories. The masks and patterns produced as software are used to recognise otherwise well known picture patterns, as in the case of optical correlators.

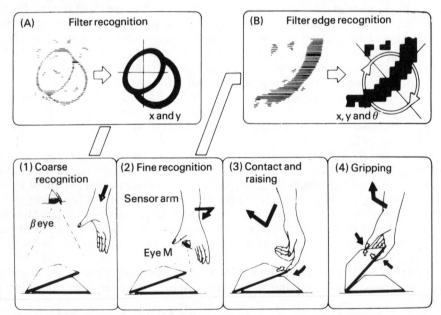

Figure 6.35   Filter recognition and pick-up in vacuum cleaner assembly. (Photo: Hitachi Ltd.)

286

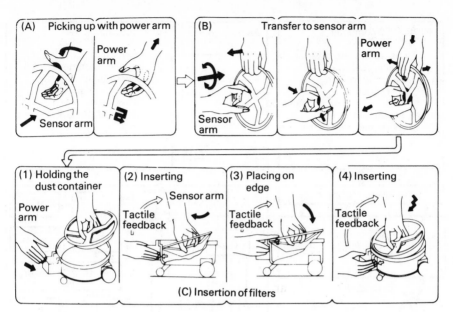

Figure 6.36   Insertion of filter during vacuum cleaner assembly. (Photo: Hitachi Ltd.)

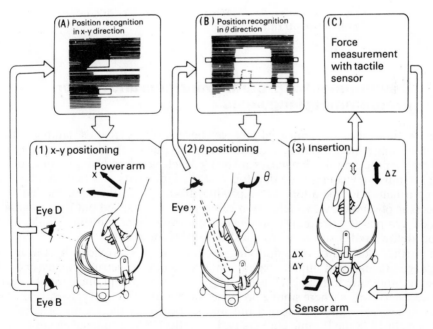

Figure 6.37   Positioning the motor by optical feedback in vacuum cleaner assembly. (Photo: Hitachi Ltd.)

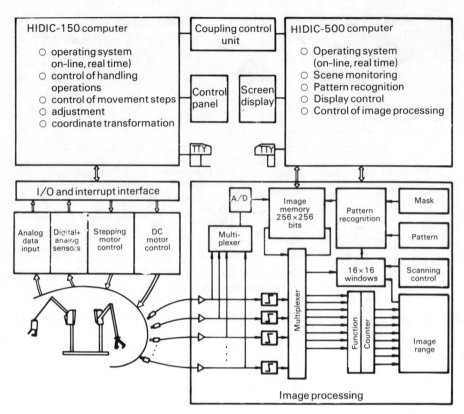

*Figure 6.38. Structure of the control system for vacuum cleaner assembly.
(Photo: Hitachi Ltd.)*

6.3 Development of programming aids and higher programming languages

The flexible use of industrial robots necessarily demands simple, rapid programming of new handling operations. For this purpose both programming aids and higher programming languages are being developed. For example, an auxiliary programming device has been developed at the University of Rhode Island, with which a handling operation is "preproduced" (6.33). In this case the handling process is observed by two television cameras, and is carried out by a manually guided gripper (Figure 6.39) provided with three light emitting diodes. The associated desired positions are established by pressing a button.

A further possibility of simplifying programming is provided by the use of language recognition units based on frequency spectrum analysers, for programming with human language [6.16]. This development, incidentally, is also of interest to clinics where simple processes such as the filling of glasses, operation of switches, etc. must be carried out automatically in the rooms of paralysed patients. In 1978 the Purdue University, Lafayette/USA, simulated an application with a model setup. The method of programming by language is also being tried out on machine tools, and suitable language recognition units, which can at present

288

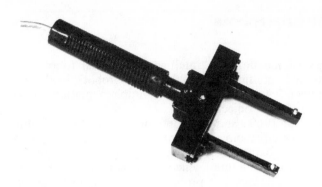

only identify an extremely limited vocabulary, have been on the market for some years.

In addition to these programming aids, work is being done increasingly on higher programming languages which enable programming to be carried out and errors to be eliminated easily and quickly, even by unskilled persons. For this purpose several developments are in progress, all of which aim to be compared, as far as programming convenience is concerned, with programming languages for numerically controlled machine tools such as APT, EXAPT and the like. Most attention is being focussed on language developments for automatic assembly with industrial robots.

One of the most advanced programming languages is the AL language (Assembly Language) [6.34] similar to ALGOL, developed at Stanford University, Menlo Park/USA. The following brief example for picking up a box and placing it on a pallet characterises the efficiency of this language:

MOVE ARM TO box;
CLOSE FINGERS, ON GRASP SENSOR DO STOP:
IF FINGERS < 25* INCHES THEN BEGIN {error} ... END:
AFFIX box TO ARM:
MOVE box TO pallet VIA (point 1, point 2),
ON FORCE * Z > 3 * OUNCES DO STOP:

Compared with NC programming languages, one peculiarity of this language lies in the fact that it can be used in combination with "teach-in programming". Further language developments are in progress worldwide. For example, work is being done at the University of Edinburgh, Great Britain, on the RAPT (Robot APT) programming system similar to the APT [6.35]. SRI International, Menlo Park/USA, is investigating a FORTRAN-like language, RPL (Robot Programming Language) [6.27]. IBM presented their AUTOPASS system some time ago [6.36], whilst the Artificial Intelligence Laboratory of MIT, Cambridge/USA) is launching its language LAMA [6.37]. Politechnico di Milano, Milan, are testing their BASIC-like, in FORTRAN implemented language MAL (Multipurpose Assembly Language) on their experimental carrier SUPERSIGMA [6.38].

This short list of current activities in the development of higher programming languages should not, however, conceal the fact that despite the relatively large number of attempts, no breakthrough has yet been achieved in higher programming languages for industrial robots. Nevertheless we must expect that in a few years efficient programming systems will be available for industry.

289

6.4 "Intelligent industrial robots"

We are hearing increasingly of the phrases "intelligent industrial robots" or "industrial robots with artificial intelligence". If the word "robot" is already emotive, the adjective "intelligent" will certainly not help implementation unless it is explained in greater detail and more objectively.

Since about the beginning of the 1960's and the more intensive use of electronic digital computers and their ever advancing development, scientists have been concerned with the new direction in data processing, with a special partial aspect of their discipline, which they call "artificial intelligence". For examples, computers have long been able to solve intelligence test problems based, for instance, on geometrical analogy codes [6.39]. Computers can also understand simple sentences of living languages such as English or Italian, and interpret them. Chess playing computers have also become very popular [6.40].

Explained in simplified terms, the work in artificial intelligence means to investigate and simulate human capabilities relating to the handling of objectives, the sensitive possibilities of linguistic understanding, the solving of problems, and finally the ability to plan one's own action. All these facilities are characteristics of the feature known colloquially as "intelligence". If the term is analysed closely, the essence of intelligence can be defined generally as follows [6.41]:

1. Construction of a picture of the outside world, which is constantly being improved by learning,
2. the ability to make a suitable selection and association of information, formation of invariables and their storage.
3. construction of algorithms of behaviour and the checking of these algorithms for their practical application by playing them through on a internal model of the outside world,
4. construction of algorithms for evaluating such algorithms and the ability to replace algorithms which have proved unsuitable or can no longer be adapted to a changed environment by better ones,
5. anticipation of future situations in the outside world by their simulation on the internal model.

This factual description of intelligence characteristics enables an answer to be given to the question of which and in what form these characteristics can be transferred to technical and – human created – artificial systems.

The presence of sensors such as those described for the field of handling engineering in Section 6.1, is a primary precondition for the construction of images. But just as the production of a photograph requires no intelligence from the camera, these sensors do not represent intelligent systems in themselves. The presence of intelligence is characterised only by the ability to process information and to create other types of information from this on the basis of a learning ability. However, learning or learning ability is therefore also the main precondition for the creation of artificial intelligence, and the followign definitions are the basis of technical learning processes in particular [6.42]:

1. Learning means acquiring or improving on a characteristic or capability of a system by interaction (information exchange) with an environment and, where appropriate, under the influence of a teacher, according to a target or success criterion.
2. Adaptation means the adaptation of a system to an environment, in order to optimise characteristics, capabilities or conditions of the system in this environment.

From the point of view of present development, it must be stated here that within the foreseeable future primarily adaptive industrial robots have still to be developed and then tested within a broad spectrum.

The broader intelligence characteristics described are required in the field of industrial robot applications when their use can no longer be planned to the fullest extent and when, as a result, suitable algorithms can no longer be constructed for conceivable situations and "implanted" in or communicated to the industrial robot by programming. In the industrial field, such situations are conceivable when, over a period of time, fully automatic and autonomous production is to be generated. In this case automatic machines must be capable of recognising problems and structuring them so that the appropriate, but as yet non-existent algorithms, can be constructed automatically for solving the new problem.

Such questions are already relevant to space technology when it is necessary, for example, to survey the surface of distant stars by means of portable robots. Controlling the robot from the earth control station is no longer possible here because the time for the information to be transmitted between the vehicle and the control station may be several minutes. Thus only one overriding strategy is communicated to the robot, on the basis of which its own strategies can be derived, taking into consideration the immediate conditions prevailing.

6.5   Prospects for future development

The preceding sections describe in detail the developments that have taken place in the years 1975 to 1978 in experimental facilities of larger companies and research institutions, as well as refinements of these developments. If we disregard the concrete details of these developments, since every prototype can certainly be further improved before it can be applied industrially on a broader basis, certain trends have been recognised which would indicate a wider use of industrial robots in the foreseeable future.

The widening of the application base and hence the quantitative growth of the industrial robots used, depends on two developments with a different time horizon. Firstly it is necessary to open up further fields of application for the existing industrial robots in those industries which have so far made little or hardly any use of industrial robots. This is a problem of diffusion where innovation barriers must be broken down and where new technologies are always opposed, although strictly speaking industrial robots are not "new technologies" and although an innovation resistance which cannot be underestimated is present in the word "robot". NC technology, by comparison, has never had to fight against such emotional innovation barriers, although evidence can be produced in this field that an enormous increase in productivity can be achieved with NC machine tools [6.43]. However, one additional problem, difficult to overcome with industrial robots, is the fact that their use represents from the very beginning a big jump in automation compared with the introduction of NC technology in machine tools, whilst in the case of machine tools themselves, different stages of development in mechanisation and automation overlapped. The size of this overall jump in automation is demonstrated by the fact that the use of an industrial robot, even on an NC machine tool, once again increases the degree of automation of the whole system.

A glimpse into the more remote future can only be speculative, but long-term technology forecasts which have been made in the past have shown that although the prediction of a technological stage of development cannot apply with any accuracy to a particular point in time, the general statements made may well apply.

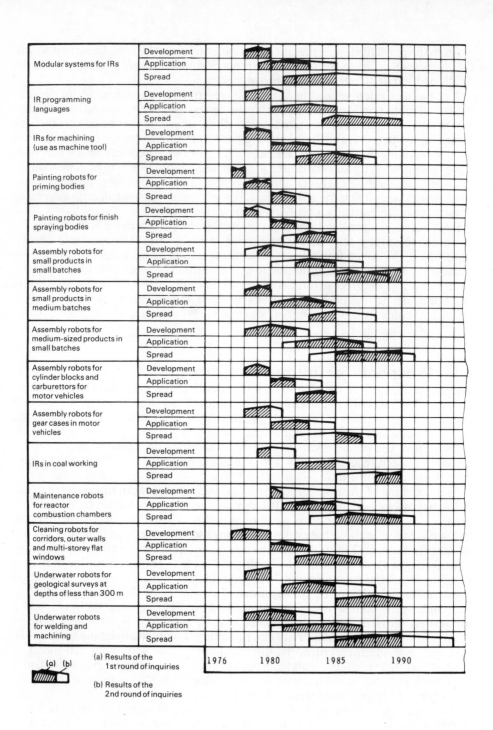

Figure 6.40   Delphi forecast for the technical development of industrial robots
              (according to [6.44]).

292

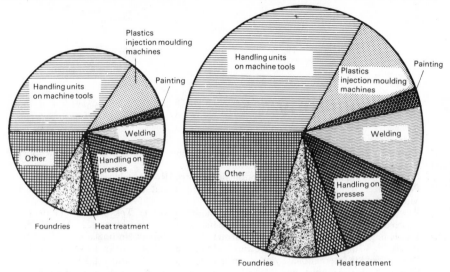

1980	1985
Total turnover: 167.5 US$ also: Auxiliary equipment and periphery: 58.4 US$	Total turnover: 398.9 US$ also: Auxiliary equipment and periphery: 119.6 US$

Figure 6.41   *Forecast of market development for industrial robots and pick-and-place units in Japan (according to [6.44]).*

Against this background of a sceptical evaluation of long-term technological forecasts, however, the results of an expert inquiry conducted in Japan in 1976, in the form of a Delphi forecast on the further development of industrial robot technology, are worthy of note [6.44]. Some results of this forecast are reproduced in Figure 6.40.

The purely quantitative long-term development is seen in Japan against the numerical values contained in Figure 6.41. In this case, however, it must be remembered that the turnover figures relate both to simple pick-and-place units and to industrial robots already on the market. According to this sales forecast, the application areas are shifting only slightly in relation to each other. Comparable estimates for Sweden [6.45] assume that the number of pick-and-place units and industrial robots in use will double every three years. It is therefore calculated that in 1978 about 600 units (not only industrial robots!) were used, and that these figures will increase in 1980 to 1200 and by 1985 to 5000.

The forecast relating to the quantitative use of sensor guided industrial robots, e.g. for assembly and inspection, is far more cautious by comparison. The extent of application still very largely depends here on the successes achieved in further technological development. For Japan, however, a developing market amounting to billions (measured in US$) is not considered impossible in this field.

7. Bibliography

[1.1] Warnecke, H. J.; Schraft, R. D.: Industrieroboter. (Industrial robots.) Mainz: Krausskopf 1973.

[1.2] Young, J. F.: Robotics. London: Butterworths, 1973.

[1.3] Proceedings 2nd International Symposium on Industrial Robots, Chicago, 16 – 18.5.1972.

[1.4] Proceedings 3rd International Symposium on Industrial Robots, Zurich, 29 – 31.5.1973, Munich: Verlag moderne Industrie, 1973.

[1.5] Proceedings 4th International Symposium on Industrial Robots, Tokyo, 19 – 21.11.1974.

[1.6] Proceedings 5th International Symposium on Industrial Robots, Chicago, 22 – 24.9.1975.

[1.7] Proceedings 6th International Symposium on Industrial Robots, Nottingham, 24 – 26.3.1976.

[1.8] Proceedings 7th International Symposium on Industrial Robots, Tokyo, 19 – 21.10.1977.

[1.9] Proceedings 8th International Symposium on Industrial Robots, Stuttgart, 30.5 – 1.6.1978.

[1.10] Steinbuch, K.: Automat und Mensch. (Automatic machine and man.) Berlin/Göttingen/ Heidelberg: Springer, 1963.

[1.11] Foith, J. P.: Szenenanalyse aus der Sicht natürlicher und künstlicher Intelligenz. (Scene analysis from the point of view of natural and artificial intelligence.) Unpublished report of the Institute for Information Processing in Engineering and Biology (IITB), Karlsruhe, 1976.

[1.12] König, M.: Aufgaben und Lösungsmöglichkeiten für Sensoren von Industrierobotern. (Tasks and possible solutions for sensors of industrial robots). Vortragsmappe 5. IPA Arbeitstagung (Working Meeting), Stuttgart, 21 – 23.1.1975.

[1.13] Norman, D. A.: Models for Human Memory in: Encyclopedia of Linguistics, Information and Control. Oxford: Pergamon Press, 1967, pp. 311-313.

[1.14] Got, T. a.o.: Compact Packaging by Robot with Tactile Sensors, in (5), pp. 149-159.

[1.15] VDI-Guideline 3239: Symbols for Feed Functions. Definitions, symbols, code numbers, application [10.66].

[1.16] VDI Guideline 3240: Feeding Devices. Examples [10.71].

[1.17] Heer, E.: Remotely Manned Systems. Pasadena: California Institute of Technology, 1973.

[2.1] Klemenz, D.: Programmierbare Fluidik-Steuerungen VII. (Programmable fluidics control systems.) 7th Cranfield Fluidics Conference, Stuttgart 1975, Paper 4.

[2.2] Klemenz, D.: Programmierbare pneumatische Steuerungen. (Programmable pneumatic control systems.) Industrieanzeiger 99 (1977) No. 25, pp. 436-439.

[2.3] Klemenz, D.: Pneumatisch/fluidische Verknüpfungsmatrix für programmierbare Steuerungen. (Pneumatic/fluidic association matrix for programmable control system.) HGF-Report 77/64.

[2.4] Sinning, H.: CNC-Steuerung für Industrieroboter mit mehreren Bewegungsachsen. (CNC control system for industrial robots with several axes of motion.) Zeitschrift für wirtschaftliche Fertigung 72 (1977) No. 4, pp. 177-181.

[2.5] Geisselmann, H.: Fernsehsensor und seine Verkettung mit einen Industrieroboter. (The television sensor and its connection to an industrial robot.) 8th International Symposium on Industrial Robots, Stuttgart, 1978.

[2.6] Paul, R.: WAVE – A Model Based Language For Manipulator Control. The Industrial Robot 4 (1977) No. 1, p. 10.

[2.7] Bernorio, M., Bertonic, M., Solmalvico, M.: Programming A Robot in Quasi-Natural Language. The Industrial Robot 4 (1977) No. 3, p. 132.

[2.8] Popplestone, R. J., Ambler, A. P., Bellos, I.: RAPT – A Language For Describing Assemblies. The Industrial Robot 5 (1978) No. 3, p. 131.

[2.9] Wallner, J. M.: Das MTM System als Rationalisierungs- und Kalkulationsgrundlage. (The MTM system as a rationalisation and calculation basis.) Verlag Technische Rundschau Berne and Stuttgart.

[2.10] VDI 3239: Symbols for feed functions. Beuth-Vertrieb GmbH, Berlin and Cologne, 1966.

[2.11] Unpublished report of the Arbeitsgemeinschaft Handhabungssysteme – Handling Systems Working Group – (Working Committee on Grippers), 1978.

[2.12] Auer, B. H.: Beitrag zur Steigerung der Flexibilität von Handhabungseinrichtungen im Bereich der Einzel – und Kleinserienfertigung und Montage. (Contribution to the increase in the flexibility of handling devices in the field of single and small batch production and assembly.) Berlin 1977, Dissertation.

[2.13] Kreis, W.: Greifer für Warmbetriebe; Fördern und Heben. (Grippers for high temperature operations.) Fördern und Heben, 27 (1977), No. 1 Fachteil mht.

[2.14] Wauer, G., Greller, P.: Schnelles Anpassen von Greifern an unregelemässigen Werkstücken mit Abformbacken. (Rapid adaptation of grippers to irregular workpieces with formed jaws.) Fördern und Heben, 26 (1976), no. 13, Fachteil mht.

[2.15] Macri, G. C.: Analysis of first installation failures; technical paper, SME, MS 77-735, 1977.

[2.16] Hermann, G.: Analyse von Handhabungsvorgängen im Hinblick auf deren Anforderungen an programmierbare Handhabungsgeräte in det Teilefertigung. (Analysis of handling processes with regard to their requirements for programmable handling equipment in parts production.) Stuttgart 1976, Dissertation.

[2.17] Schraft, R. D., Haaf, D.: Möglichkeiten der automatischen Montage in der Kleinserienfertigung. (Possibilities of automatic assembly in small batch production.) Proceedings 8th ISIR; Stuttgart 1978.

[2.18] Okada, T., Tsuchiya, S.: On a versatile finger system. Proceedings 7th ISIR Tokyo 1977, pp. 345-352.

[2.19] Hanfusa, H., Asada, H.: Stable prehension by a robot hand with elastic fingers. Proceedings 7th ISIR, Tokyo 1977, pp. 361-367.

[2.20] Cardaun, V.: Greifer für Industrieroboter. (Grippers for industrial robots). Fördern und Heben. Fachteil Montage- und Handhabungstechnik. (Specialist section on assembly and handling engineering.) 28 (1978), no. 1, pp. 40-43.

[2.21] Spur, G., Furgac, I., Rall, K.: Entwicklung einer flexiblen Fertigungszelle mit integriertem Handhabungsgerät. (Development of a flexible production cell with integrated handling unit.) Proceedings 8th ISIR/4th CIRT, 10th IPA Working Meeting 1978.

[3.1] VDI Guideline 3254: Numerically controlled machine tools. Accuracy specifications. Berlin, Cologne: Beuth-Vertrieb 1971.

[3.2] Engel, G.: Messverfahren zur Analyse des dynamischen und kinematischen Verhaltens von Handhabungsgeräten und Industrierobotern. (Measurement method for the analysis of the dynamic and kinematic behaviour of handling units and industrial robots.) Industrieanzeiger (1976) No. 99, pp. 1774-1777.

[3.3] Weck, M., Miesen, W., Engel, G., D'Souza, C.: Experimental Analysis of Handling Devices and Industrial Robots. Proceedings CIRP-General Assembly, Paris 1978.

[3.4] VDI Guideline 2058: Assessment of working noise at the workplace with regard to hearing damage. Berlin, Cologne: Beuth Vertrieb 1971.

[3.5] DIN 45641: Averaging level and assessment level for temporally fluctuating sound processes. Berlin, Cologne: Beuth Vertieb 1974.

[3.6] 7th Occupational Disease Decree of 20.6.1968. Appendix 1, Number 26: Hardness of hearing and Deafness due to Noise.

[3.7] Brodbeck, B.: Wirtschaftlichkeitsuntersuchung bei der Automatisierung det Handhabung am Beispiel eines IR-Einsatzes. (Cost effective study for the automation of a handling process, taking the example of an industrial robot application.) Montage- und handhabungstechnik (mht) (1976) No. 3, pp. 117-119.

[3.8] Schimke, E. F.: Planung von Handhabungssystemen. (Planning of handling systems.) Dr.-Ing. Dissertation Technical University of Aachen, 1976.

[3.9] Schiefelbusch, H.: Wirschaftlichkeitsfragen zum Einsatz von Industrierobotern. (Economic questions relating to the use of industrial robots.) Fördern und heben (Fachteil mht) 1977) No. 10, pp. 78-81.

[3.10] Brodbeck, B.: Untersuchung des Arbeitsverhaltens programmierbarer Handhabungs-geräte. (Study of the operating behaviour of programmable handling devices). Dr.-Ing. Diss. University of Stuttgart. Book Series "IPA Forschung und Praxis", Krasskopf-Verlag Mainz 1979.

[4.1] Warnecke, H. J., Weiss, K.: Katalog Zubringeeinrichtungen. (Catalogue of Feeding Devices.) Book series "Produktionstechnik heute". Krausskopf-Verlag Mainz 1978.

[4.2] Herrmann, G. Analyse von Handhabungsvorgängen im Hinblick auf deren Anforderungen an programmierbare Handhabungsgeräte in der Teilefertigung (Analysis of handling processes with regard to their requirements for programmable handling units in parts production.) Dissertation University of Stuttgart 1976).

[4.3] Frank, H. E.: Handhabungseinrichtungen. (Handling Devices.) Book series: "Produktionstechnik heute". Krausskopf-Verlag Mainz 1975.

295

[4.4] Württ. Ingenieurverein Stuttgart (publishers): Werkstückhandhabung in der automatisierten Fertigung. (Workpiece handling in automated production.) Ein Leitfaden zur Lehrschau. Stuttgart 1968.

[4.5] Hesse, S., Zapf, H.: Verkettungseinrichtungen in der Fertigungstechnik. (Interlinking devices in production). C. Hanser-Verlag Munich 1971.

[4.6] König, M.: Aufgaben und Lösungsmöglichkeiten für Sensoren von Industrie-robotern. "Er fahrungsaustausch über Industrieroboter". (Tasks and possible solutions for sensors of industrial robots. "Exchange of experiences on industrial robots".) 5th IPA Working Meeting.

[4.7] Graf, B., Knappmann, R., Schmidt, J., Weiss, K.: Automatischer flexibler Bohrarbeitsplatz für Kleinserien. (Automatic, flexible drilling workplace for small batches.) Technische Rundschau No. 8 and No. 10, 1979.

[5.1] Herrmann, G.: Analyse von Handhabungsvorgängen im Hinblick auf deren Anforderungen an programmierbare Handhabungsgeräte in der Teilefertigung (see 4.2). Stuttgart; Dr.-Ing. Dissertation 1976.

[5.2] Ropohl, G.: Flexible Fertigungssysteme. (Flexible production systems.) Mainz: Krausskopf-Verlag 1971.

[5.3] Herrmann, G.: in: Neue Handhabungssysteme als technische Hilfen für den Arbeitsprozess. (New handling systems as technical aids for the working process.) Special report on research projects of the same name initiated by the Federal Ministry of Research and Technology. December 1975.

[5.4] Schraft, R. D.: Systematisches Auswählen und Konzipieren von programmgesteuerten Handhabungsgeräten. (Systematic selection and design of program-controlled handling devices.) Mainz: Krausskopf-Verlag 1977.

[5.5] Analytische Arbeitsbewertung für sie Metallindustrie in Nordwürttemberg/Nordbaden. (Analytical job evaluation for the metal industry in Northern Württemberg/North Baden.) Verband Württ.-Badischer Metallindustrieller, Stuttgart 1967.

[5.6] Schmidt-Steier, U., Weiss, K.: Rechnerunterstützte Einsatzplanung von Industrierobotern und peripheren Einrichtungen in der Werkstückhandhabung. (Computer-assisted application planning of industrial robots and peripheral devices in workpiece handling). Proceedings 4th Conference of Industrial Robot Technology & 8th International Symposium on Industrial Robots, Böblingen, 30.4-1.5.1978.

[5.7] Technical Aids for the Working Process. Concluding Report on Research Projects of the Same Name initiated by the Federal Ministry of Research and Technology (BMFT), November 1973.

[5.8] Warnecke, H. J., Weiss, K.: Katalog Zubringeeinrichtungen – Hilfsmittel zur Planung von Handhabungssystemen. (Catalogue of feeding devices – Aids to planning handling systems.) Mainz: Krausskopf-Verlag 1978.

[5.9] Konold, P., Kern, H., Reger, H.: Arbeitssystem Elemente Katalog – Hilfsmittel zur Planung von Arbeitssystemen. (Catalogue of Working System Elements – Aids to Planning working Systems.) Mainz: Krausskopf-Verlag 1977.

[5.10] Zangenmeister, C.: Grundzüge der Nutzwertanalyse in der Systemtechnik. (Basic principles of value analysis.) Munich: Wittemannsche Buchhandlung 1970.

[5.11] Schmidt-Streier, U.: Planung des Industrieroboter-Einsatzes mit Hilfe der EDV. (Planning of industrial robot application by means of EDP). f + h – fördern und heben 27 (1977), Fachteil mht.

[5.12] Warnecke, H. J., Schraft, R. D., Schmidt-Streier, U.: Computer Graphics Planning of Industrial Robot Application. Proceedings 3rd CISM IFTOMM International Symposium on Theory and Practice of Robots and Manipulators, Udine, 12-15.9.1978.

[5.13] Schimke, E-F.: Planung und Einsatz von Industrierobotern. (Planning and application of industrial robots). Düsseldorf: VDI-Verlag 1978.

[5.14] Gnedenko, B. W., Kowalenko, I. N.: Einführung in die Bedientheorie. (Introduction to Control and Operating Theory). Munich: Oldenbourg Verlag 1971).

[5.15] Schöne, A.: Simulation technischer Systeme. (Simulation of technical systems) – Volume 3. Munich: Hanser Verlag 1974.

[5.16] Warnecke, H. J., Schraft, R. D.: Katalog Industrieroboter. (Catalogue of industrial robots.) Mainz: Krausskopf-Verlag (in preparation).

[5.17] Groh, W.: Das Ordnen von Massenteilen und ihre selbsttätige Zuführung in die Werkzeugmaschine. (The orientation of mass produced parts and their automatic feeding into machine tools). Werkstatttechnik u. Maschinenbau 47 (1957) No. 8.

[5.18] Frank, E.: Werkstückverhalten – Ausgangspunkt für die Automatisierung der Werkstückhandhabung. (Workpiece behaviour – basis for automatic workpiece handling.) Technische Rundschau (1972) No. 22.

[5.19] Bronner, A.: Vereinfachte Wirschaftlichkeitsrechnung. (Simplified cost effectiveness calculation). Braunschweig: Vieweg-Verlag 1971.

[5.20] Blohm, H., Lüder, K.: Investition. (Investment.) Munich: Vahlen-Verlag 1972.

296

[5.21] Brodbeck, B.: Wirtschaftlichkeitsuntersuchung bei der Automatisierung der Handhabung am Beispiel eins Industrieroboter-Einsatzes. (Cost effectiveness study in the automation of handling taking the example of an industrial robot application). Montage- und Handhabungstechnik (mht), 3/1975.

[5.22] Hagekötter, M., a.o.: Bemerkungen und Thesen zum Arbeitsschutz. (Remarks and theses on Working Safety), Federal Institute for Working Safety (Labour Welfare) and Accident Research. Dortmund 1974.

[5.23] Schulz, U.: Statistik als Grundlage der Unfallforschung. (Statistics as a basis for accident research.) Main Union of Industrial Employers' Liability Insurance Associations. Bonn 1973.

[5.24] South German Refined and Unrefined Metal Employers' Liability Insurance Association, Technical Year Reports 1968/69 to 1972/73. Stuttgart.

[6.1] Schweizer, M.: Taktile Sensoren für programmierbare Handhabungsgeräte. (Tactile sensors for programmable handling devices.). Dr.-Ing. Dissertation University of Stuttgart 1978.

[6.2] Martin, R.: Mittlere Technologie ist flexibel. (Average technology is flexible.) VDI News No. 22, 1978.

[6.3] Minoru, U.: Tactile Sensors for Industrial Robots to Detect a Slip. Proceedings of the 2nd Symposium on Industrial Robots, Chicago, 1972.

[6.4] Köhler, W.: Sensoren für Industrieroboter. (Sensors for industrial robots.) VDI News No. 18, 1978.

[6.5] Hirt, D., Isenberg, G., Krovinovic, Z.: Entwicklung eines adaptiven freisystems. (Development of an adaptive grasp system.) Fördern und heben (Fachteil mht) 26 (1976) No. 13, pp. 24-26.

[6.6] Goto, T: Precise Insert Operation by Tactile Controlled Robot. Proceedings of the 4th International Symposium on Industrial Robots, Tokyo, 1974. pp. 378-386.

[6.7] Watson, P. C.: A Multidimensional System Analysis of the Assembly Process as Performed by a Manipulator. First Annual North American Robot Conference October 28th, Chicago, Illinois, 1976.

[6.8] Salmon, M.: Das Sigma-System zur Automatisierung von Montage- und Bearbeitungsvorgängen. (The Sigma system for automatic assembly and machining processes). Fördern und heben (Fachteil mht) 26 (1976) no. 3, p. 109.

[6.9] Nevins, J. L., Whitney, D. E.: Adaptable-Progammable Assembly Systems: An Information and Control Problem. Proceedings of the 5th International Symposium on Industrial Robots. Chicago 1975.

[6.10] Proceedings of the 7th Symposium on Industrial Robots, Tokyo 1977.

[6.11] Schweizer, M.: Sensoren für programmgesteuerte Handhabungsgeräte. (Sensors for program-controlled handlings units.) Fördern und heben (Fachteil mht) 27 (1977) no. 4, pp. 22-27.

[6.12] Baker, H. H.: Machine Vision in an Industrial Environment 6th Int. Symp. on Ind. Rob. March 1976, Nottingham.

[6.13] Bretschi, J., König, M.: Leistungsfähige Sensoren für Industrieroboter. (Efficient sensors for industrial robots.) Fördern und heben (Fachteil mht) 26 (1976) No. 13, pp. 42-44.

[6.14] Zambuto, D. A., Chaney, J. E.: An Industrial Robot with Mini-Computer-Control. GTE Laboratories Inc., USA, Proceedings of the 3rd Conf. on Ind. Rob. Technol. and 6th Int. Symp. on Ind. Rob. March 1976, Nottingham.

[6.15] Rosen, C. A.: Material Handling Robots for Programmable Automation. Proceedings of the 1st IFAC Symposium on Manufacturing Technology, Tokyo, 1977, pp. 147-152.

[6.16] Nitzan, D., Rosen, C. A.: Programmable Industrial Automation. IEEE Transactions on Computers, vol. C-25 (1976) No. 12, pp. 1259-1270.

[6.17] Brodbeck, B.: Industrieroboter in Japan. (Industrial robots in Japan). Fördern und heben (Fachteil mht) 28 (1978) No. 1, pp. 33-36.

[6.18] Gonsalves, R. A., Kurlat, S.: Robotics in Japan from an American Perspective. SME Technical Paper MS 77-754.

[6.20] Rosen, C. A., Nitzan, D.: Use of Sensors in Programmable Automation. Computer (1977) No. 12, pp. 12-23.

[6.21] Heginbotham, W. B., Pugh, A.: Roboter mit Sehvermögen. (Robots with visual ability.) Lecture No. 31 of the 5th IPA Working Meeting, Stuttgart, 1975.

[6.22] Birk, J., Kelley, R., a.o.: General Methods to Enable Robots with Vision to Acquire, Orient and Transport Workpieces. 4th report GRANT APR 74 – 13935, University of Rhode Island, Kingston, R.J., 1978.

[6.23] Tsuboi, Y., Inoure, T.: Robot Assembly System Using TV-Camera. Proceedings of the 3rd Conf. on Ind. Rob. Techn. and 6th ISIR, Nottingham, 1976.

[6.24] Nevins, J. L., Whitney, D. E.: Computer-Controlled Assembly. Scientific American 238 (1978) No. 2, pp. 62-74.

297

[6.25] Schraft, R. D., Haaf, D.: Möglichkeiten der Automatischen Montage in der Kleinserienfertigung. (Possibilities of automatic assembly in small batch production.) Proceedings of the 8th ISIR, Stuttgart, 1978.

[6.26] Exploratory Research in Advanced Automation. Second Report 1974 GRANT GJ 381 X 1. National Science Foundation, Washington, 1974.

[6.27] Rosen, C. A., Nitzan, D.: Machine Intelligence Research Applied to Industrial Automation. 8th Report GRANT APR 75 – 13074. SRI International, Menlo Park, Cal., 1978.

[6.28] Wang, S. M., Will, P. M.: Sensors for Computer.Control Mechanical Assembly. Industrial Robot 5 (1978) No. 1, pp. 9-16.

[6.29] Bedini, R. a.o.: Academic Research on the Olivetti SIGMA-System Applications. Proceedings of the 75h ISIR, Tokyo, 1977.

[6.30] Nakano, E.: Cooperational Control of the Anthropomorphous Manipulator MELARM. Proceedings of the 4th ISIR, Tokyo, 1977.

[6.31] Takeyasu, K. a.o.: An Approach to the Integrated Intelligent Robot with Multiple Sensory Feedback: Construction and Control Functions. Proceedings of the 75h ISIR, Tokyo, 1977.

[6.32] Kashioka, S. a.o.: An Approach to the Integrated Intelligent Robot with Multiple Sensory Feedback: Visual Recognition Techniques. Proceedings of the 7th ISIR, Tokyo, 1977.

[6.33] Birk, J. R., Kelley, R. B.: New Robot Programming Devices for Teaching Assembly, Inspection, Materials Handling, and Palletizing Tasks. Proceedings of the 6th ISIR, Nottingham, 1976.

[6.34] Binford, T. O.: Computer Integrated Assembly Systems. Proceedings 6th NSF Grantees' Conf. on Prod. Research, West Lafayette, 1978.

[6.35] Popplestone, R. J., a.o.: RAPT: A Language for Describing Assemblies. Industrial Robot 5 (1978) No. 3, pp. 131-137.

[6.36] Lieberman, L. J., Wesley, M. A.: AUTOPASS: An Automatic Programming System for Computer Controlled Mechanical Assembly. IBM Journ. of Res. and Dev. (1977) No. 7.

[6.37] Lozano-Perez, T.: The Design of a Mechanical Assembly System. MIT Artificial Intelligence Lab., Techn. Report 397. MIT, Cambridge, 1976.

[6.38] Colombetti, M. a.o.: Manipulating by High Level Languages. Proceedings Conf. on Applied Robotics, Pilsen, 1977.

[6.39] Winston, P. H.: Artificial Intelligence. Reading, Menlo Park, London, Amsterdam: Addision-Wesley 1977.

[6.40] Michie, D.: New Face of A.I. Experimental Programming Reports No. 33. Machine Intelligence Research Unit, University of Edinburgh, 1977.

[6.41] Klaus, G.: Wörterbuch der Kybernetik. (Dictionary of Cybernetics.) Frankfurt, Hamburg: Fischer 1969.

[6.42] Peipmann, R.: Erkennen von Strukturen und Mustern. (Recognition of structures and patterns.) Berlin, New York; de Gruyter 1976.

[6.43] Fewer, More Productive Machines. American Machinist 122 (1978) No. 12, pp. 133-148.

[6.44] Yonemoto, K., Shiino, K.: Present State and Future Outlook of Technology and Market on Industrial Robots in Japan. Proceedings 7th ISIR, Tokyo, 1977.

[6.45] Over 600 Industrial Robots in Sweden. VDI-N. 32 (1978) No. 45, p. 48.